THE IRISH AMERICAN
FAMILY
ALBUM

THE IRISH AMERICAN FAMILY ALBUM

DOROTHY AND THOMAS HOOBLER

INTRODUCTION BY JOSEPH P. KENNEDY II

OXFORD UNIVERSITY PRESS • NEW YORK • OXFORD

Oxford University Press

Oxford   New York
Athens   Auckland   Bangkok   Bombay
Calcutta   Cape Town   Dar es Salaam   Delhi
Florence   Hong Kong   Istanbul   Karachi
Kuala Lumpur   Madras   Madrid   Melbourne
Mexico City   Nairobi   Paris   Singapore
Taipei   Tokyo   Toronto

and associated companies in
Berlin   Ibadan

Copyright © 1995 by Dorothy and Thomas Hoobler
Introduction copyright © 1995 by Oxford University Press, Inc.

Design: Sandy Kaufman
Layout: Greg Wozney
Consultant: Hasia Diner, Professor of American Studies,
            University of Maryland at College Park

Published by Oxford University Press, Inc.,
200 Madison Avenue, New York, New York 10016

Oxford is a registered trademark of Oxford University Press, Inc.

Library of Congress Cataloging-in-Publication Data

Hoobler, Dorothy.
     The Irish American family album / Dorothy and Thomas Hoobler;
     introduction by Joseph P. Kennedy II.
     p. cm. — (American family albums)
     Includes bibliographical references and index.
     1. Irish Americans—History—Juvenile literature.
     [1. Irish Americans—History.]
     I. Hoobler, Thomas. II. Title. III. Series.
E184.I6H66   1995

973'.049162—dc20   94-19569
                                                            CIP
                                                            AC

ISBN 0-19-508127-7 (lib. ed.); ISBN 0-19-509461-1(trade ed.); ISBN 0-19-509125-6 (series, lib. ed.)

9 8 7 6 5 4 3 2

Printed in the United States of America
on acid-free paper

Cover: The McGovern family of Philadelphia, 1904.

Frontispiece: Businessman and farmer Joseph Powell and his family traveled to Savannah from
Emanuel County, Georgia, for this studio portrait in 1908.

Contents page: Hattie Finnegan of Idaho, around 1900.

CONTENTS

On September 24, 1952, Joseph P. Kennedy II was born in Brighton, Massachusetts. As the son of Senator Robert F. Kennedy (who served as U.S. attorney general from 1961 to 1964) and nephew of former President John F. Kennedy, he has carried on his family's political traditions and continued the fight for human rights.

A graduate of the University of Massachusetts at Boston, Joseph Kennedy went on to head the Citizens Energy Corporation, a nonprofit energy assistance company. Under his guidance, the company provided low-cost heating oil and prescription drugs for the elderly and the poor.

In 1986, Joe Kennedy won election as a U.S. representative from the 8th District of Massachusetts, the same district that his great-grandfather John "Honey Fitz" Fitzgerald had represented from 1895 to 1901 and that John Kennedy had represented from 1949 to 1953.

Since winning election to Congress, Joe Kennedy has worked to improve health care coverage for working people, war veterans, and the elderly, as well as create affordable housing, stimulate job growth, and end discrimination in the banking industry. He is currently a member of the House Banking, Finance, and Urban Affairs Committee as well as the Veterans Affairs Committee.

Kennedy's commitment to human rights has taken him to many countries, including Northern Ireland, Haiti, Germany, and Armenia, to speak out against oppression. He is married to Beth Kelly Kennedy and is the father of twin boys.

A Kennedy family photo taken in July 1938 at the U.S. Embassy in London. Ambassador Joseph Kennedy and his wife, Rose, are in the center, surrounded by their children (from left to right): Eunice, John, Rosemary, Mary Jean, Edward, Joseph, Jr., Patricia, Robert, and Kathleen.

Robert Kennedy and his wife, Ethel, with their children at Hickory Hill estate in McLean, Virginia, 1967. Joseph Kennedy II is standing next to his father.

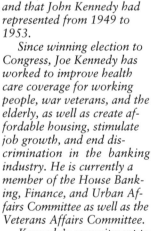

In Boston, Kennedy visits patients at a veterans' hospital.

Congressman Joe Kennedy with his wife, Beth, in 1993.

INTRODUCTION

by U.S. Representative Joseph P. Kennedy II

In the middle of the 19th century, eight young immigrants from different parts of Ireland set out in the space of a few months for the promise of a new start in America. The country they left behind was ravaged with blight, famine, and disease. The ocean crossing, in crowded ships over perilous seas, was long and dangerous.

Their destination, Boston, was hardly the promised land of the immigrant imagination. The penniless new arrivals found themselves in a city inhospitable to their religion, their language, and their culture.

But bolstered by family and faith, the eight immigrants, my great-great-grandparents, managed to secure a foothold in the new world, joining thousands of earlier arrivals in the crowded neighborhoods of their new homeland.

One of those ancestors was Patrick Kennedy, from New Ross, County Wexford. He became a cooper—a barrel maker—on the wharves of East Boston. More than a century after his arrival in America, his great-grandson, President John F. Kennedy, my uncle, returned to New Ross to visit the rolling green coastland that Patrick Kennedy had left behind.

Standing on the dock from which his great-grandfather had embarked on the long journey, President Kennedy said that Patrick Kennedy had "carried nothing with him except two things: a strong religious faith and a strong desire for liberty. I am glad to say that all his great-grandchildren have valued that inheritance."

That legacy, as strong as ever, has always included a commitment to helping those left behind, the poor and the weak and the sick. Patrick Kennedy's son, Patrick Joseph Kennedy, entered ward politics in East Boston as a way of assuring jobs and services for the needy people of his community.

John "Honey Fitz" Francis Fitzgerald, my great-grandfather, came up through a similar path in the North End of Boston before going on to become a congressman and mayor of Boston.

Public service, a way of protecting the community, came to mean for John Fitzgerald a way of helping the broader community of man. The memory of the indignities suffered by his parents' generation in a city where "No Irish Need Apply" signs were common moved him to help other ethnic groups under attack.

One of his toughest battles after going to Washington was standing up against the anti-immigrant tide then sweeping the country. "It is fashionable today to cry out against the immigration of the Hungarian, the Italian and the Jew," he said in a speech on the floor of the House of Representatives in 1897, "but I think that the man who comes to this country for the first time—to a strange land without friends and without employment—is born of the stuff that is bound to make good citizens."

The enduring lesson of the Irish immigrant experience is the belief in the commonality of man—that anyone from anywhere, given a chance, can forge a new start in America, and that the poor and the dispossessed already here can no more escape their condition without help than my immigrant forebears did.

My father, Robert F. Kennedy, devoted his career as attorney general under President Kennedy, and later, as a U.S. senator from New York, to carrying out the ideals of equal opportunity—ending discrimination in housing, education, and employment and ensuring the voting rights of all Americans.

On the day Dr. Martin Luther King, Jr., was killed, my father was campaigning for the Presidency in Indianapolis, where he delivered the terrible news to a crowd at a city park. His statement that night urging calm and reconciliation conveys the vision of America as a place where people are ready and willing to live and work together.

"What we need in the United States is not division," he said. "What we need in the United States is not hatred. What we need in the United States is not violence or lawlessness, but love and wisdom, and compassion toward one another, and a feeling of justice towards those who still suffer within our country, whether they be white or they be black."

In my own career as congressman and founder of a nonprofit energy company, I have tried to carry on the legacy of compassion and cooperation.

America pulling together—every color, creed, and faith, individual citizens, private corporations and government agencies—is the legacy born of the struggling immigrant experience, an inheritance that belongs to us all.

Joe Kennedy

The Mulhern family of Philadelphia kept this picture of "Grandmom McShane" with some of her family members in Ireland in 1912.

CHAPTER ONE

THE OULD SOD

The Irish have long been great storytellers. In the days when few people could read or write, wandering bards sang the history of Ireland to avid listeners in every village in the land. The tales are full of heroes and villains, saints and sinners, bravery and betrayal. Knowledge of their history has helped the Irish people keep alive their culture through centuries of persecution. The harp that the bards played is still one of the symbols of Ireland.

A little more than 2,000 years ago, people known as Celts arrived in Ireland from the European mainland. They spoke Gaelic, which became the national tongue of the Irish people. Over time, five Celtic kingdoms arose—Ulster, Leinster, Munster, Connaught, and Meath. According to tradition, one of the chiefs of the five kingdoms received the title of "high king," bestowed at a hill called Tara.

The Irish Celts frequently crossed the Irish Sea to attack the neighboring island of Britain. In the 5th century, one of these raiding parties brought back a teenaged English slave named Patrick. After escaping from his master, Patrick fled Ireland. But around the year 432, after becoming a priest, he returned to bring Christianity to the Irish people. Using the three leaves of the Irish shamrock to explain the Christian doctrine of the trinity (in which a single God consists of three persons—the Father, Son, and Holy Spirit), Saint Patrick created one of the most beloved Irish symbols.

Later Christian missionaries established monasteries that became famous for their beautiful, hand-copied manuscripts. Women, too, devoted their lives to Christ, joining communities of nuns. One was established by Saint Bridget of Kildare, whose name became as popular at the baptism of Irish girls as Patrick's was for boys.

According to legend, an Irish missionary named Brendan sailed westward into the Atlantic, where he discovered a land inhabited by people he had never heard of. For centuries after that, seafarers looked for "St. Brendan's Isle." Though it was never found, some modern scholars wonder if Brendan might have been the first Irishman to reach America.

In the year 795, seafaring warriors known as the Vikings swept down from their Scandinavian homeland into Ireland. During the next two centuries, the Vikings built forts along the east coast. One of these grew into the city of Dublin, which would become the capital of the modern nation.

For about a century, the Vikings' control of the east remained unchallenged. Then Brian Boru, a king of Munster, took the title of high king and organized an army to drive the Vikings out. Brian's forces won a decisive victory at the Battle of Clontarf in 1014.

Weakened by the Viking invasions, Ireland became prey to another invader—the Normans, who had already conquered England in 1066. Around 1155, the Norman king Henry II was granted title to Ireland by Pope Adrian IV (the only English pope). Henry launched an invasion of Ireland in 1169. The Irish resisted, but before long the English had established a domain in eastern Ireland, called the Pale.

Henry and his successors granted Irish lands to Norman lords. These Norman families, such as the Fitzgeralds, would in time become as Irish as the older Celtic clans. But beyond the Pale, Celtic chiefs (called "barbarians" by the English) retained their control.

Over time, the English cut down the vast forests of Ireland for use in shipbuilding. In the centuries since then, Irish have gathered fuel for their fires from the peat bogs that are found throughout the country. The "ould sod," as the Irish came to call their land, was the source of food and warmth.

Until the 16th century, religion played no part in the conflict between Irish and English, for both groups were Christians loyal to the pope. In 1534, however, King Henry VIII of England broke away from the Catholic church and established the Church of England, with himself as its head. The Catholic Irish—Norman and Celtic families alike—resisted the attempts of Henry and his daughter Elizabeth I to force Protestantism on Ireland.

In 1603 James I succeeded Elizabeth. Because James's wife was Catholic, the Irish hoped he would end their religious persecution. However, James's hold on the throne was shaky; to keep the support of Scottish Presbyterians, he allowed them to settle in Ulster, the northern province of Ireland. This move had tragic and long-lasting results, for Ulster remains separated from the Catholic south even today.

In 1649 James's son Charles I was overthrown by a rebellion that made Oliver Cromwell the lord protector of England. Cromwell, an ardent Protestant, ruthlessly tried to stamp out Catholicism in Ireland. His troops slaughtered men, women, and children; destroyed churches; and hunted down priests. Cromwell granted large areas of Irish land to his supporters, thus beginning a system in which English landlords owned the farms on which Irish tenants toiled.

When the English monarchy was restored under the Stuart fam-

ily in 1669, Irish Catholics who had fought in support of the Stuarts expected to reap the rewards. But by now, the English people had become thoroughly Protestant. The English Parliament deposed James II, the second Stuart king, and in 1688 offered the throne to his Protestant son-in-law, William of Orange.

James II fled to Ireland, where

Members of a large Irish family in front of their stone hut with thatched roof in the 1880s. Many Irish lived their whole life in the same house in which they were born.

Catholics rallied to his defense. William pursued him with an army, and on July 12, 1690, defeated James's forces at the Battle of the River Boyne—a defeat that still rankles in the hearts of Irish patriots. Ever since, Irish Protestants have been called the "men of Orange," and the wearing of green has symbolized Irish independence. Having established control over all of Ireland, the English set up an Irish parliament. It passed a series of "penal laws" that restricted the rights of the vast majority of Irish who clung to their Catholic religion. Catholics were unable to

hold public office, vote, become lawyers, buy land, own guns, or even own a horse worth more than five pounds. The use of the Gaelic language was discouraged. According to one historian, the English regarded Irish Catholics as the equivalent of farm animals—necessary only to work the land for their English landlords.

The Irish stubbornly resisted this attempt to stamp out their culture and faith. With their churches closed, Catholics attended masses in fields and forests. Gaelic remained the common language of the people, and Irish poets kept the literary tradition alive.

The English domination affected even the Scots-Irish Presbyterians in the north. When they began to develop a cloth-making industry, the English Parliament passed a law requiring all Irish-made cloth to be shipped to England and then instituted high tariffs, or import taxes, to make the trade unprofitable. Such moves served only to bring Presbyterians and Catholics together in a common cause.

In 1791 Wolfe Tone, a Dublin Protestant, helped found a group called United Irishmen, which aimed to form an Irish government of both Catholics and Protestants. Driven into exile by the British, Tone went to France, which was at war with England, to gain support for his cause.

In 1798 two Irish rebellions broke out: one in Ulster (led by Protestants) and the other in

County Wexford on the east coast (led by a Catholic priest, Father John Murphy). The British brutally crushed the uprisings, arresting thousands of people and torturing them mercilessly to obtain the names of other rebels. By the time Tone arrived with a force of French soldiers, the Irish rebels had been defeated. Tone was captured, and he committed suicide in prison.

The English tried to pacify the Irish by passing the Act of Union of 1801. This act officially united Ireland with Britain; the Irish Parliament was dissolved, and Irish of all religions were permitted to elect delegates to the British Parliament in London. Though many of the penal laws were repealed, some restrictions on Catholics remained. They could vote in elections but could not serve in Parliament. In addition, Catholics had to pay a tithe—10 percent of their income—to support the Protestant Church of Ireland.

In 1803 the United Irish patriots staged their last attempt at resistance. Robert Emmett, a veteran of the 1798 rebellion, led an uprising in Dublin. Captured and sentenced to death, Emmett gave a speech on the gallows that was remembered wherever Irish gathered: "Let no man write my epitaph.... When my country takes her place among the nations of the earth, then, and not till then let my epitaph be written."

A new leader now made his appearance, aiming to win "eman-

cipation" or full rights, for Ireland's Catholic majority. He was Daniel O'Connell, a Catholic landowner who attracted followers with his spellbinding oratory. O'Connell formed a political rights association that charged members the "Catholic rent" of a penny a month. Though O'Connell won election to Parliament in 1828, he was denied a seat because he refused to take the required oath of

An Irish woman patiently knits while waiting for a meal to cook. This photograph, from around 1920, reflects the poverty in which many families still lived long after the famine.

opposition to Catholicism. Forced to speak from the visitors' gallery, he eloquently defended the Irish cause. The following year, the English granted Irish emancipation.

Meanwhile, O'Connell set another goal: dissolving the union and making Ireland a separate nation. O'Connell rallied his supporters at huge outdoor meetings, including one at Tara in 1843 that attracted half a million people.

As O'Connell aged, younger and more radical leaders took up the cause. A group called Young Ireland defiantly glorified Irish heroes of the past, especially the

rebels of 1798. In 1848, a year after O'Connell's death, another rebellion broke out, but it, too, was crushed.

By that time, Ireland had begun to experience the greatest catastrophe in its history: the Great Famine. Between 1845 and the mid-1850s, a blight destroyed much of Ireland's potato crop. Potatoes had been the staple diet of the Irish farm families, and the crop failures caused mass starvation throughout the country.

Rather than try to feed the starving farmers, the English landlords began to evict them from their humble, sod-roofed cottages for nonpayment of rent. It became profitable for the landlords to clear the land of hostile and starving Irish people so that it could be used for grazing farm animals.

It is estimated that a million Irish died during the famine. Millions more rebelled in the only way they could—by emigrating. The scope of the tragedy becomes clear from the population figures. In 1841 more than 8 million people lived in Ireland; by 1900 the population had dropped to less than half of that. Starvation, disease, and emigration accounted for the loss, from which Ireland has never recovered.

The Irish who came to the United States did not forget the proud history of their homeland. With their support, a successful Irish revolution would take place in the 20th century—though the fight for a united, independent Ireland continues today.

IRELAND

The shaded area on this map of Ireland shows the six northern counties of Ulster that are still part of Great Britain. Ulster, Leinster, Connacht, and Munster were the names of four ancient Celtic kingdoms.

An illustration from a London newspaper in 1849 shows an Irish mother and children trying to find potatoes that were not affected by the crop blight.

FAMINE

During the 1840s a virus infected the potato crops throughout Ireland, causing widespread starvation. An Irish priest, Father Theobald Matthew, described the beginnings of the Great Famine in 1845.

On the 27th [of July], I passed from Cork to Dublin, and this doomed plant bloomed in all the luxuriance of an abundant harvest. Returning on the third [of August], I beheld with sorrow one wide waste of putrefying vegetation. In many places the wretched people were seated on the fences of their decaying gardens, wringing their hands and wailing bitterly the destruction that had left them foodless.

Diarmuid O'Donovan Rossa lived through the famine as a child. Although many farmers grew grain and other crops besides potatoes, those went to pay the rent. Rossa remembered:

Coming on the harvest time of the year 1845, the crops looked splendid. But one fine morning in July there was a cry around that some blight had struck the potato stalks. The leaves had been blighted, and from being green, parts of them were turned black and brown, and when these parts were felt between the fingers they'd crumble into ashes. The air was laden with a sickly odor of decay, as if the hand of Death had stricken the potato field, and that everything growing in it was rotting. This is the recollection that remains in my mind of what I felt in our marsh field that morning, when I went with my father and mother to see it.

The stalks withered away day by day. Yet the potatoes had grown to a fairly large size. But the seed of decay and death had been planted in them too. They were dug and put into a pit in the field. By and by an alarming rumor ran through the country that the potatoes were rotting in the pits. Our pit was opened, and there, sure enough, were some of the biggest of the potatoes half rotten. The ones that were not touched with the rot were separated from the rotting ones, and were carted into the "chamber" house, back of our dwelling house. That chamberhouse had been specially prepared for them, the walls of it being padded with straw, but it was soon found that the potatoes were rotting in the chamber too. Then all hands were set to work to make another picking; the potatoes that were rotting were thrown in to the back yard, and these that were whole and appeared sound were taken up into the loft over our kitchen. The loft had been specially propped to bear the extra weight. But the potatoes rotted in the loft also, and before many weeks the blight had eaten up the supply that was to last the family for the whole year.

Then one of our fields had a crop of wheat, and when the wheat was reaped and stacked, the landlord put "keepers" [guards] on it, and on all that we had, and these keepers remained in the house till the wheat was threshed and bagged, and taken to the mill. I well remember one of the keepers going with my mother to Lloyd's mill, just across the road from the marsh field and from the mill to the agent, who was in town...that day, to receive rents.

When my mother came home she came without any money. The rent was £18 a year. The wheat was thirty shillings a bag; there were twelve bags and a few stone, that came in all to £18, 5s., and she gave all to the agent.

After a bitter winter, the blight reappeared a second year. The magistrate of Cork, Nicholas Cummins, toured the country in 1846 and described the conditions in a letter.

In the first [house that I visited], six famished and ghastly skeletons, to all appearances dead, were huddled in a corner on some filthy straw.... I approached with horror, and found by a low moaning they were alive. They were in fever, four children, a woman, and what had once been a man.... In a few minutes, I was surrounded by at least 2000 such phantoms, such frightful spectres as no words can describe.... Their demoniac yells are still ringing in my ears, and their horrible images are fixed on my brain. My heart sickens at the recital, but I must go on....

I found myself grasped by a woman with an infant just born in her arms and the remains of a filthy sack across her loins—the sole covering of herself and baby. The same morning, the police opened a house...which was observed shut for many days, and two frozen corpses were found, lying upon the mud floor, half devoured by rats.... In another house, within 500 yards of the cavalry station at Skibbereen, the dispensary doctor found seven wretches lying unable to move, under the same cloak. One had been dead many hours, but the others were unable to move either themselves or the corpse.

The famine caused by the potato blight weakened the Irish people and made them more susceptible to disease. A parish priest described the suffering and death at the height of the famine.

I was called in to prepare a poor fellow, whose mother lay beside him dead two days.... I was called in two days after to a miserable object, beside whom a child lay dead, for the twenty-four hours previous; two others lay beside her just expiring, and, horrible to relate, a famished cat got upon the bed, and was just about to gnaw the corpse of the deceased infant, until I prevented it.... A third of the population has been already carried away. Every morning four or five corpses are to be found on the street dead, the victims of famine and disease.

In 1847 Ireland had suffered famine for two years. W. E. Forster, a relief worker for the Society of Friends, described the horror:

The town of Westport was in itself a strange and fearful sight, like what we read of in beleaguered cities; its streets crowded with gaunt wanderers, sauntering to and fro with a hopeless air and hunger-struck look; a mob of starved, almost naked, women around the poorhouse, clamoring for soup tickets.

One poor woman whose cabin I had visited [in a rural district] said, "There will be nothing for us but to lie down and die." I tried to give her hope of English aid, but, alas! her prophecy has been too true. Out of a population of 240 I found thirteen already dead from want. The survivors were like walking skeletons—the men gaunt and haggard, stamped with the livid mark of hunger—the children crying with pain—the women in some of the cabins too weak to stand.

A group of Irish children during the mid-19th century, when the famine had devastated all parts of the country.

Bridget Fitzgerald, who came to the United States in 1921 as a girl of 17, described living under British rule:

Yeah, England didn't do anything for Ireland. They took everything away from the Irish and give the Irish nothing. If you were a Roman Catholic you didn't stand a chance in Ireland. There were no Catholic schools, but they had Protestant schools and good teachers and seven or eight children going to school and two teachers. That's how it was. I think the British were afraid to educate the Irish.

It was up to the Catholics themselves to get an education. What could they do? I went to school three years and that's about it. Some education. I didn't realize it then, but now I know. I didn't have the mind to think about those things when I was young. If I had, God knows what I would be today. I'd be a rebel of some kind.

An outdoor mass in County Donegal in 1867. By that time the British government had repealed the laws barring the practice of the Roman Catholic religion, but many Irish Catholic communities could not afford to build a church.

Rosalie B. Hart Priour, who immigrated to Texas in 1834, related how her grandfather, a Catholic convert, had been tortured during the rebellion of 1789.

He was arrested and whipped in the streets of Dublin for three days in succession. The executioner, after laying on so many strokes, would stop and ask: "Will you renounce the Romanish Church and tell where the priests are hidden? If you will, you will...receive a reward from the government of England." His answer was invariably, "No! I joined the Catholic Church because I thought it was the true religion, and I will not dishonor myself, or sell my soul for any worldly advantages that this world can confer."

My dear grandmother remained by his side during the whole of that trying scene, with my father then an infant, six months old, in her arms...and at every blow the flesh and blood of my dear grandfather would fall over them, but the executioner seemed heartless, and resisted all her entreaties until the third day when he gave her possession of my grandfather with permission to take him home and save him if she could, but it was too late. With all the care lavished on him he died, another martyr for the Catholic religion.

After Bridget O'Donnel immigrated to America, she recalled being evicted from her house in western Ireland in 1849, at the height of the Great Famine. Her husband was gone, she was pregnant, and she had two small children to care for.

Dan Sheedey and five or six men came to tumble [knock down] my house; they wanted me to give [up] possession. I said that I would not; I had fever, and was within two months of my down-lying [giving birth]; they commenced knocking down the house, and had half of it knocked down when two neighbours, women, Nell Spellesley and Kate How, carried me out. I had the priest and a doctor to attend to me shortly after. Father Meehan anointed me [gave her the sacrament for the dying]. I was carried into a cabin, and lay there for eight days, when I had the creature born dead. I lay for three weeks after that. The whole of my family got the fever, and one boy of thirteen years old died with want and with hunger while we were lying sick. Dan Sheedey and Blake took the corn into Kilrush, and sold it. I don't know what they got for it. I had not a bit for my children to eat when they took it from me.

Paul O'Dwyer, who would become a prominent lawyer in the United States, began his autobiography with these words:

My birth in 1907 was not regarded as a blessing in my family. I was the eleventh and last O'Dwyer in an already overcrowded five-room house, and while I know my deeply religious mother proclaimed me to be a gift from heaven, I doubt that the other members of the family subscribed to her view.

We lived in Lismirrane, a seventeen-house country village which was part of the parish of Bohola....

Bohola is about in the center of County Mayo, in the west of Ireland, Gaelic in tradition and name. *Mayo* means "plain of the yew tree." Centuries ago invaders, first Normans and then English, found that the woods gave refuge to attacking rebels, so the forest was cut down. There were no yew trees by the time I was born, and very few of any others. Mayo had been overpopulated for many centuries. It has supplied more immigrants to the world than any other area of equal size, with the possible exception of certain parts of Sicily. At different times the landlords tried to drive small farmer-tenants off the land in order to make Mayo a great grazing preserve. The natives' resistance took many forms, the least militant of which was ostracism. The treatment Mayo farmers gave to an English land agent, Captain Boycott, for his tyrannical eviction of tenants, supplied another word to their acquired language.

Though Catherine Moran McNamara was over 90 when she was interviewed in 1976, her memories of conditions in Ireland during her childhood remained vivid.

There was twelve in our family. The oldest died and the other one went to Australia with my uncle. I was about five when she went. So there was ten of us, you might say, in our family. We had to pay *every cent* we possibly could produce to taxes. Every war England had, she had you pay her part, even though you just had nothing, and you had to pay on your land some expenses out of it....

The Irish lived under awful stress. I've seen [another] family thrown out. I recall that distinctly because we took them in our barn.... I seen the little child, this is God's truth, I'll never forget this, it was just about a year and a half, put out in the little cradle. I seen the pots put out and the coals of fire put into the iron oven they used to bake with. Everything that they had, put into the yard. If they were caught in that yard that night they'd be shot or somethin'.

England did this, of course, and her regime. She had certain ones to do it. The landlord, he was English, and the English owned Ireland then....

This was goin' down six hundred years—imagine being under anyone's thumb for that length of time! But one consolation, when America opened up. It took an awful lot of needy people here and it opened a gap for them, like God done at sea, otherwise they had to be on their knees for England.

Daniel O'Connell

In 1843 Daniel O'Connell announced a "monster meeting" at Clontarf, a town outside Dublin. Here, eight centuries earlier, the Irish high king Brian Boru had defeated the Vikings. The purpose of the meeting was to urge the repeal of the Act of Union of 1801, which had incorporated Ireland into the United Kingdom.

Born in County Kerry in 1775, Daniel O'Connell went to France and England for an education. Returning to Ireland in 1798, he saw the terrible reprisals that the British took against the defeated United Irish rebels. The experience made him a determined opponent of any form of violence.

O'Connell began to organize the Catholic Association, a grass-roots movement, to petition for Catholic rights. Financed by small contributions from people throughout the country, his movement rapidly gained widespread support. A generation of Irish leaders, some of whom later came to the United States, gained their organizing experience in the Catholic Association.

O'Connell ran for a seat in Parliament, even though, as a Catholic, he could not serve. His victory by a wide margin persuaded the British to pass the Catholic Emancipation Act, which allowed Catholics to vote and hold office.

When O'Connell called the meeting at Clontarf, he hoped that a million people would gather to show their support. But British soldiers arrived to block the crowds who flocked to the meeting. O'Connell prevented bloodshed by urging his followers to disperse. Nevertheless, the British imprisoned him.

To many younger Irish, the lesson was clear: O'Connell's peaceful methods would not work. By the time O'Connell was released from jail, others had taken control of the Irish independence movement, abandoning O'Connell's refusal to use violence. But since his death in 1847, the Irish people have acclaimed him as one of their liberators.

As Others Saw Them

The African American leader Frederick Douglass was in Ireland in early 1845, just before the famine. He described the poverty he saw in Dublin:

The spectacle that affected me most, and made the most vivid impression on my mind, of the extreme poverty and wretchedness of the poor of Dublin, was the frequency with which I met little children in the street at a late hour of the night, covered with filthy rags, and seated upon cold stone steps, or in corners, leaning against brick walls, fast asleep, with none to look upon them, none to care for them.... During my stay in Dublin, I took occasion to visit the huts of the poor in its vicinity—and of all places to witness human misery, ignorance, degradation, filth and wretchedness, an Irish hut is pre-eminent.... Four mud walls about six feet high, occupying a space of ground about ten feet square, covered or thatched with straw...without floor, without windows, and sometimes without a chimney—a piece of pine board laid on the top of a box or an old chest—a pile of straw covered with dirty garments...a picture representing the crucifixion of Christ, pasted on the most conspicuous space on the wall—a few broken dishes stuck up in a corner...a man and his wife and five children, and a pig.... Here you have an Irish hut or cabin, such as millions of the people of Ireland live in. And some live in worse than these. Men and women, married or single, old and young, lie down together, in much the same degradation as the American slaves.

GROWING UP IN IRELAND

Ann McNabb, a cook, described her life in Ireland in the 19th century before she emigrated. Her American employer, a writer, took down her story in Irish dialect.

I was born nigh to Limavaddy; it's a pretty town close to Londonderry. We lived in a peat cabin, but it had a good thatched roof. Mother put on that roof. It isn't a woman's work, but she was able for it.

There were sivin childher of us. John an' Matthew they went to Australia. Mother was layin' by for five year to get their passage money. They went into the bush. We heard twice from thim and then no more. Not another word and that is forty year gone now—on account of them not reading and writing. Learning isn't cheap in them old countries as it is here, you see. I suppose they're dead now—John would be ninety now—and in heaven. They were honest men. My mother sent Joseph to Londonderry to larn the weaver's trade. My father he never was a steddy worker. He took to the drink early in life. My mother an' me an' Tilly we worked in the field for Squire Varney. Yes, plowin' an' seedin' and diggin'—any farm work he'd give us. We did men's work, but we didn't get men's pay. No, of course not. In winter we did lace work for a merchant in Londonderry. (Ann still can embroider beautifully.) It was

Nora Joyce, an Irish American immigrant, kept this photo of her sister Sarah's family in front of their cottage in Ireland.

The effects of the famine continued for decades. In the 1880s, when this photo was taken, about one out of every ten Irish men and women left their homeland for the United States. Relatives who had emigrated earlier often sent money to help pay for their ship tickets.

pleasanter nor diggin' after my hands was fit for it. But it took two weeks every year to clean and soften my hands for the needle.

Pay was very small and the twins—that was Maria and Philip—they were too young to work at all. What did we eat? Well, just potatoes. On Sundays, once a month, we'd maybe have bit of flitch [bacon]. When the potatoes rotted—that was the hard times! Oh, yes, I mind the famine years. An' the corn-meal that the 'Mericans sent. The folks said they'd rather starve nor eat it. We didn't know how to cook it. Here I eat corn dodgers and fried mush fast enough.

Maria—she was one of the twins—she died the famine year of the typhus and—well, she sickened of the herbs and roots we eat—we had no potatoes.

Bridget Fitzgerald, who came to the United States in 1921, described her early life in Ireland.

My mother would've had about fourteen children, or fifteen, sixteen, maybe. There was ten that lived. Over there they have a flock of children, and the older one watches the next one, the next one watches the next one, and the next one watches the next one; that's how they do it. If the mother would have to wait on every one, well, what would it be like? The parents don't worry about one child. My mother used to take an old wooden tub—we didn't have a bathroom, you know—and she'd fill it up full of water and she'd throw the kids in it all at once and one washed the other. And that's how we washed. And then the next three kids would do it.

When we were children you got your own food, you went looking for it. If you didn't have it in the house, you went out on the farm and got it. You had turnips, you had parsnips, you had parsley, you had scallions, you had lettuce. So we'd eat the turnips, get a trout with a pin and a piece of thread, build a

John Lane, a dairy farmer from Mount-collins, County Limerick, on his way to the creamery.

Saint Patrick

According to legend, the Christian missionary Patrick climbed the Mount of Crom in County Mayo sometime in the 5th century A.D. The mountain, home of the pagan god Crom, was the holiest spot in the pagan religion of the Celtic Irish. For 40 days and nights, Patrick wrestled with demons on the mountaintop. Some appeared in the form of snakes, and Patrick drove them out of the country.

Victorious, Patrick descended and dedicated the mountain to his god. Ever since, Catholics have made pilgrimages to the site, today called Croagh Patrick. Modern pilgrims flock there on the last Sunday in July, known as Black Crom's Sunday.

The man who is the patron saint of Ireland was born in Britain, the son of a deacon of the Catholic church. When Patrick was 16, Irish raiders carried him off to their island as a slave. For six years, he tended his master's sheep, but kept his devotion to Christianity. After receiving a sign from God in a dream, he escaped to Britain.

But, according to his own account, the "voice of the Irish" would not leave Patrick. He decided to return to preach the true faith. Boldly, Patrick carried on a series of debates with the Druids, Celtic pagan priests. Patrick's eloquence won converts to Christianity. In Downpatrick in County Armagh today stands a church built on the site of one that Patrick founded. The saint's bones are said to be buried there.

The day of his death, March 17, is celebrated as St. Patrick's Day by Catholics. All over the world, the Irish have made it their special holiday. St. Patrick's Day parades have wound through the streets of American cities since colonial times. The largest takes place in New York City, which is now home to more people of Irish descent than any other city in the world. Each March 17th, as the gala parade passes by the New York cathedral named for the saint, the "wearin' of the green" links Irish Americans to their origins.

fire, get a potato, roast potatoes, cook a fish, eat raw turnips, and anything we'd eat came out of the ground. There was no canned goods.

We had our own sheep. We'd clip the sheep and take the wool off them, you know, and then we'd card it and toll it and we'd spin it. You learned from your mother and the people around you. We learned to card and spin. The older people, they knew all about the dyes, and we'd go and gather them off a rock or stone. And you'd get them different colors and boil them down. You'd boil them down in pots. We used water from the river. Then we'd put the shanks of wool and tie them and dip them one brown and the other green and whatever colors we wanted. If we wanted to leave it the natural color, we would leave it that way. We'd sent it to the weaver and she'd made blankets out of it.

My mother baked all the bread. She cooked over a fireplace, open hearth, in a big pot with a crane on it. Burned turf. We took to the bogs and dug it up. You'd dig with a two-cornered spade and you put it down on the ground and you throw it up. And you let it dry a little bit, and then, when it's been about two weeks laying there, you go and turn it over. When you turn it over and you dry it some more, then you stack it. It makes a wonderful fire, beautiful. Beautiful. It smells wonderful.

We used to pull flax. You pull it out of the ground and you tie it in bundles. And they have a place where you soak it till it gets good and rotten water, and you weight it down with a stone. Then pull it up on the riverbank and dry it. And the kids all go out and gather it up and bundle it. And they ship it to England and England doesn't give you anything for it....

We lived with my grandfather and he had a farm, and while he was living that was fine. We had plenty of everything. But once he went, everything went. The government seized part of our farm for back taxes then. And a woman bought up the rest of it, and left us in a hole. We all went to work. I was nine. I hired out on a farm, herding cows, doing that kind of work. I had to live with the family.

Michael Kinney, who came to the United States from Ireland in 1930, recalled his youth.

I was raised on a farm in County Kerry. Well, you were never finished. There was no such thing as six hours or seven hours or eight hours. You got up in the morning, you start working, you keep working until it got dark, I guess. You had to milk the cows, feed the pigs, feed the chickens, the hogs. We worked every day there—Saturday, Sunday, Monday, Tuesday, Wednesday.

Francis Hackett, who became a journalist in the United States, recalled being told about hell by his mother when he was a child in Ireland.

I f you die in sin, Francis," she had then explained, "and die before you can make an act of perfect contrition, you go straight to Hell."

"Is that bad, Mud?" I inquired.

"It is terrible," said my mother. "The sinners are plunged in fire."

That sounded bad, certainly. "How deep are they plunged?" I ventured to ask.

"Enough to burn them," she said, compressing her lips.

"Up to their knees?"

"Oh, higher up than that."

"Up to their middle?"

"Yes, about up to their middle."

"But what kind of fire is it?"

"Fire and brimstone," my mother answered. "It never goes out. It goes on for eternity."

I began to be sorry for sinners, but their plight puzzled me. "Why don't they burn up?" I asked. Everything else did, especially chops.

"They can't," she said decisively. "They burn and burn and they can't burn up."

I was silenced. I was searching for a constructive hypothesis, and then it came to me. "Are they made of asbestos?"

Mary Kenny told an interviewer about her childhood in Ireland.

M y birthdate is July 21, 1938. I was born in Kilkenny, the Republic of Ireland.

Kilkenny has a population of about fifty-two thousand. I was the second of nine children, the oldest girl. I had

Mary Buckley, a two-year-old child in Ireland around 1926. Her mother died young, and Mary was raised by her grandparents.

Irish families who could afford to feed a cow enjoyed the milk, cream, and cheese that are staples of Irish cooking. The cart next to this woman indicates that she will sell the extra milk in town.

A farm family in front of their home in Mountcollins, County Limerick.

four brothers. Nine children wasn't that large a family in those days; most people had between six and twelve. We just somehow seemed to manage beautifully. Maybe we didn't have a lot of material things, but there was something else—love. We had a closeness that maybe children from smaller families don't always have.

My father was in the insurance business. In those days the insurance man had to cycle out in the country to collect the premiums every two weeks. So my father was home quite a bit. He did a lot of his work from the house.

We also had what I guess you would call a huckster shop, a small shop. We sold groceries—sugar, butter and stuff like that. We all had to help out in the store and also at home. And of course we didn't have a washing machine, dryer, dishwasher, or any of the conveniences I have gotten used to over here.

Food was simple. We grew a lot of our own, like potatoes and vegetables, and my father's sister was married to a farmer so we used to get bacon and stuff from them. She also used to make butter and bring us some. They were more affluent than we were, there were only four in that family. My father's sister helped out quite a bit....

My father died of cancer when I was eighteen.... My mother was expecting my youngest brother. Because my father was well insured we were all right for a couple of years, but things got difficult. My mother tried to make ends meet and put the others through school and everything. This necessitated my helping even more at home. We did have this little shop and that helped....

Of the nine of us, seven finished high school. Two went on

A group of children outside a row of houses in an Irish village around 1900. The thatched roofs, cut from the sod of the land, are still sprouting new grass.

to college. All the boys went to the Irish Christian Brothers Academy. All the girls went to the Presentation Convent. I was the first. I went from the first to the eleventh grade. I took a commercial course in high school; this trained me for my first job, which I got when I was about seventeen. When I was in school the emphasis was on educating the boys. The girls were trained to be good housewives, mothers. At the end of my years in school we were just beginning to realize that girls should have a little education too. My father was all for education. Maybe it was because his parents were schoolteachers.

My father handled the financial end of things. He paid most of the bills. But I would say my mother was the stronger of the two. My father was very quiet, so my mother handled a lot more than would normally be so. My mother, for example, was the disciplinarian in our house. I don't think I remember my father ever striking me or doing anything of this sort. He would threaten us with a switch, but he never followed through. My mother, however, was a different story. We didn't get away with anything.

Also in those days, they allowed you to use the paddle in schools. If you didn't do your homework or whatever, you got six of the best. The brothers were worse than the nuns. I can remember the boys—especially my older brother, who is a schoolteacher himself today—coming home with blisters on their hands.

Being the oldest, I was one of the first to help the family by working. I started in a furniture store working for my aunt and uncle. My pay was four dollars a week, out of which was deducted taxes and insurance. I brought home about three dollars. My mother usually got two and I got one.

George Birmingham wrote for the newspaper Irishmen All *in 1913 on the importance of the woman's contribution to the family income. It was traditional for the wife to take care of the family finances:*

Small wonder also that she is calm, that no joy has any power to stir her now, nor any sorrow is fierce enough to break her heart.... There were nights when she sat anxious in the kitchen, waiting for her husband to return from the fair, knowing that when he did come he would be thick of speech, blear-eyed, staggering in his gait; yet never so drunk but that he has his money safe, the price of the beasts he had sold; never so incapable but that he had sense enough to hand the greasy notes to her. And from then on they were hers; she kept them, spent them, saved them for rent, or hoarded them for a daughter's portion. James spent what custom ordained that he should spend at the fair, drank luck to a buyer, stood treat to a friend. Afterwards he asked no money of her. She clothed him; when the time came had the money by her for the rent and to meet the calls of the tax-gatherers. He never doubted that she was the manager of the money he earned.

In County Antrim in the 19th century, a barefoot woman tidies up her yard with a broom made of sticks. Shoes were a luxury for the desperately poor Irish families.

Around 1890, a group of Irish
immigrants crowd onto the
deck of a ship bound for a new
life in America.

CHAPTER TWO

GOING TO AMERICA

n the years after 1850, when millions of Irish left their homeland for the United States, an "American wake" was held on the night before their departure. Normally, a wake is a gathering around the body of a dead person. Immigrants to America were like the dead, in that no one ever expected to see them again.

At an American wake, people sang and danced to fiddles all night. Some prayed the rosary while others drank Irish whiskey and told old stories that brought memories of happier times. In the morning, the emigrants boarded a cart with their few possessions, taking a chunk of turf as a memory of the "ould sod" and a Cross of Saint Brigid made from reeds.

Despite the forced gaiety, the leave-taking was at heart a sorrowful occasion. In Gaelic, emigration was known as *deoraí* or *dibeartach*—words that mean "exile" or "banished people."

The Irish had journeyed to the English colonies in America before 1600. Some arrived as soldiers in the British army. The English, seeking to settle the territories they had gained in the New World, also encouraged Irish immigration as a source of labor.

Many of the first Irish came as indentured servants—they pledged to work for a period of years to repay the cost of their passage. Sometimes, when groups of these indentured servants reached port in America, people looking for servants or laborers came aboard the ship and bid at auction for their contracts. The system was different from slavery, for the time of service was limited, but the indentured servants were often treated like slaves.

In the 18th century, some immigrants used the "redemptioner" system: they paid part of their passage on departure and were given a certain amount of time to find a "redeemer"—friend, relative, or employer—who would contribute the rest. If no redeemer was found, however, these immigrants could be sold into indentured servitude.

The struggle between Catholics and Protestants had its effect on emigration. Catholics fled persecution at home by going to the New World. In 1632 the English king Charles I granted Maryland to Lord Baltimore, a Catholic noble, as a haven for English Catholics. Many Irish took advantage of the opportunity to move there.

In 1703 the English government began to use "transportation" to the colonies as punishment for criminals, including Irish rebels. Protestant colonists, alarmed by this inflow of "papists" (people

loyal to the pope), passed laws limiting the number of Irish newcomers. The arrival of Irish immigrants on the ship *St. Patrick* in Boston Harbor in 1636 caused an anti-Catholic riot. In 1699 even Maryland, where Protestants were now a majority, restricted the number of Irish Catholic servants allowed entry.

However, there were still some colonies where Irish were welcomed. Sir Thomas Dongan, a Catholic of English descent born in County Kildare, served as governor of New York from 1683 to 1686 and backed the Charter of Liberties that allowed freedom of religion. In Pennsylvania, founded by the Quaker William Penn, all forms of religious belief were tolerated. It was in Philadelphia in 1699 that the first public Catholic mass in the United States took place, authorized by Mayor James Logan, an immigrant from County Armagh.

Most of the Irish immigrants in colonial times were male. Thousands of them, hostile to England from birth, rallied to the support of the American Revolution and fought in the Continental army. Their contribution was so great that the Irish Parliament declared in 1784 that "America was lost by Irish immigrants." General George Washington himself paid tribute to his Irish officers and soldiers when he made *St. Patrick* the military password on the day the British evacuated Boston—March 17, 1776.

By some estimates there were about 400,000 people of Irish ancestry in the United States in 1790. About half of these were Catholics. The rest, primarily Presbyterians, came from Ulster, and later became known as the Scots-Irish.

After the United States won its independence, Irish immigrants continued to seek opportunities here, even though many states preserved colonial anti-Catholic laws. Between 1820 and 1850, the number of Irish immigrants was greater than that of any other ethnic group.

The Great Famine that devastated Ireland in the years after 1845 created a surge of emigration. Nearly 800,000 Irish immigrants arrived in the United States from 1841 to 1850; more than 900,000 followed in the next decade. In other words, more than one out of every five people in Ireland left for the United States in that 20-year period.

The journey was long and difficult, but the Irish were literally fleeing for their lives. For many, the alternative was death—either from starvation or the diseases that spread through the famished population.

The journey often began in a horse-drawn cart or on foot, as whole families headed for the seacoast. Some set sail for America from Irish ports such as Cork. The majority crossed the Irish Sea in small boats to reach the English port of Liverpool, where merchant vessels unloaded their goods before returning to America with passengers.

Unscrupulous swindlers fleeced many families of what little savings they had. "Passage-brokers"

On board an immigrant ship around 1870, a man sells dolls. In port cities, hucksters frequently took advantage of Irish farm families who had little experience with urban life.

roamed the Irish countryside, selling ship tickets that proved to be worthless. Similar scams awaited Irish emigrants in the port cities. According to one report, about 100,000 destitute Irish men, women, and children roamed the streets of English port cities by May 1847.

The ships that carried the Irish across the Atlantic were usually cargo vessels. Both the United States and Canada exported goods such as grain, timber, and cotton to Europe. It was profitable for ship captains to take on passengers for the return trip.

However, these cargo ships had no facilities for passengers in the steerage below decks. Ventilation was poor and there were no toilets or bedding and no means of cooking food. In fact, many ships provided no food at all; the passengers had to bring everything they needed for a voyage that could take 50 to 60 days. At the American wakes, neighbors donated oatcakes baked many times so that they would remain edible, though hard, through the long voyage. The ships carried casks of water for passengers, but over the long voyage it often became foul.

According to historian Oscar Handlin, on an emigrant ship in the days of sail, between 400 and 1,000 passengers would be crammed into a space 75 feet long and 25 feet wide. Herman Melville, who had been a crew member on one such ship, described the passengers in his novel *Redburn:*

The friendless emigrants, stowed away like bales of cotton and packed like slaves in a slave ship, confined in a place that, during storm time, must be closed against both light and air, who can do no cooking nor warm so much as a cup of water. Nor is this all...passengers are cut off from the most indispensable conveniences.... We had not been at sea one week when to hold your head down the forehatch was like holding it down a suddenly opened cesspool.

In such conditions, many immigrants, already weakened by

famine, succumbed to "ship fever" (typhoid fever), which claimed thousands before they could reach America. After a prayer by the living, the dead were dropped overboard in a sack of coal so they would sink to the bottom of the sea.

The immigrants also faced the danger of storms or fire. In the year 1834 alone, at least 17 immigrant ships sank on the way from Liverpool to Canada. Irish dubbed the emigrant vessels "coffin ships."

The mass emigration did not stop when the Great Famine subsided. Having found a better life in the United States (as well as freedom from the harsh English rule), Irish immigrants carefully saved their wages to pay for the passage of their relatives in Ireland. It was common for the eldest child in a family to go first, later sending for younger brothers or sisters who would in turn contribute until the whole family had arrived.

Most of the Irish immigrants were young people; it has been said that Ireland's greatest export was its children. Even after the famine subsided, America seemed a better choice to young Irish people who wanted a better life.

Before the famine, Irish sons had shared equally in whatever inheritance their parents left. But the famine made it necessary for families to choose one son (not always the eldest) to inherit the meager family holdings. Thus, only that one son could afford to marry and start a family. His brothers and sisters could either remain as single members of the family household or emigrate.

By the 1870s oceangoing steamships had replaced sailing vessels, cutting the time of the trip across the Atlantic Ocean to about two weeks. Conditions aboard improved, too, as the United States and Great Britain passed laws

The faces of these immigrants in the 1920s appear cheerful as they gather around to hear traditional songs played on the fiddle and flute. Such activities helped to make the trip to America easier.

regulating the treatment of passengers. Because Ireland remained a relatively poor nation, the tide of immigration into the United States continued.

After 1890 Irish women immigrants outnumbered the men—the only immigrant group in which this was the case. The reasons are complex. Irish women were in demand as servants, a better-paying job than most Irish men could find. By contrast, the opportunities for women were severely restricted in Ireland. Many could not even hope to become wives and mothers, for there were relatively few men who could afford the dowry needed to marry. Irish women often chose to go to America and seek husbands there rather than remain single at home.

By the 20th century, Irish arrived on modern ocean liners that provided decent conditions even for those with the lowest-priced tickets. One woman who arrived in the 1920s remembered the trip as being like a week's vacation between hard work in Ireland and hard work in the new country.

In the century from 1830 to 1930 about 4 million Irish came to the United States— more than the entire population of the Irish Republic today. A new immigration law passed in 1924 restricted the number of newcomers who were permitted entry. During the Great Depression years of the 1930s Irish immigration dropped to 13,000 per year—about 6 percent of what it had been in the previous decade.

Irish immigration into the United States remained relatively low until about 1980. In the 1980s and 1990s, however, Ireland has suffered from severe unemployment, and the rate of emigration has picked up again.

In the U.S. census of 1990, nearly 39 million Americans listed their ancestry as Irish—more than any other ethnic group except German and roughly eight times as many people as the population of today's Ireland, North and South. It is no exaggeration to say that the United States has become the second home of the Irish people.

Often members of a family moved to the United States one at a time, with the first earning money to bring over other family members. Dominic Cassidy, born in Belfast, Ireland, in 1916 described the pattern in his family:

I was among the last of the family to come to this country. My first brother left in 1921, a little after the great war. There was no opportunity in Ireland for him to get work, so he emigrated to America along with his wife. When he landed here he sent for my sister; my sister left in 1922. They in turn sent for the rest of us; another brother in 1923, another sister in 1924. My parents and I came together in 1927. So the exodus started in 1921, and we all finally wound up here in 1927 as a family.

This cartoon, published in a U.S. magazine in 1854, shows an Irishman reading a poster that advertises ship passage to New York. His battered hat and the pack over his shoulder indicate that he may be dreaming of a change in his fortunes.

THE DECISION TO LEAVE

For the desperately poor Irish refugees from the famine, obtaining enough money to emigrate was a daunting task. Ann McNabb told how her family raised the funds.

Mother said when Maria died [during the famine], "There's a curse on ould green Ireland and we'll get out of it." So we worked an' saved for four year an' then Squire Varney helped a bit an' we sent Tilly to America. She had always more head than me. She came to Philadelphia and got a place for general housework at Mrs. Bent's. Tilly got but $2 a week, bein' a greenhorn. But she larned...to cook and bake and to wash and do up shirts—all American fashion. Then Tilly axed $3 a week. Mother always said, "Don't ax a penny more tha you're worth. But you know your own vally and ax that."

She had no expenses and laid by money enough to bring me out before the year was gone.... Me and Tilly saved till we brought Joseph and Phil over, and they went into Mr. Bent's mills as weaver and spool boy and then they saved, and we all brought out my mother and father.

The arrival of a letter from someone who had gone to America was a big event in an Irish village. It might contain an "American Ticket," a money order that was a welcome sight to a poverty-stricken family. A schoolteacher in Donegal, who was often called on to read such letters to illiterate families, described the excitement they caused.

All members of the family and some friends gather round. Each one gives his opinion as to the best way of opening the letter without damaging the contents. "The Scholar" is sent for. He perhaps is not at home just then, or he may be from one cause or another rather long in coming, but as everyone could distinguish an "American Ticket" by the picture of the Eagle from another part of the document, anxiety overcomes patience, and a breach is made in the cover with a scissors or knife. Not a breath escapes while this operation is being performed. Soon fold after fold of the enclosed manuscript is opened. The last one is being turned up and...the "Ticket" appears.... This part of the letter, the pearl as may be called—is taken in charge by the old woman who opens her long cloth purse...and in the innermost...cavity the American Eagle finds a safe retreat.

The letters that Irish immigrants wrote home lured others to make the trip. Margaret McCarthy wrote after arriving in New York in 1850:

My dear father.... Any man or woman are fools that would not venture to come to this plentyful Country where no man or woman ever hungered or ever will.

An Irish American ranch hand near Puget Sound in Washington Territory (not yet a state), wrote to his brother:

I still think I am in as good a country as there is in the world today for a poor man.... Any man here that will work and save his earnings, and make use of his brains can grow rich.

After graduating from high school in Ireland early in the 20th century, Francis Hackett was told that his father, a doctor, could not afford to send him to college. Hackett decided to emigrate. He recalled years later:

One day my father said, "Darby White [the town pharmacist] wants you to look in on him."... It was a somber day. He was a formidable man, with that hard hat of his, and his trim beard. He talked into his beard. "I'd like you to have something, Master Francis, since your good father tells me you are thinking of voyaging across the ocean, and there will be little things you'll want."

He pulled out a drawer, and scooped up a number of coins. These he placed on his desk in two little pillars.

"Times are hard, Master Francis, and I can't be of as much help as I would wish. But it is yours, my dear boy, with my blessing." An involuntary sigh came from him as he wheeled to me. "Now, hold out your hand." And he put the heavy gold coins into it.

"Goodbye, Master Francis," and he gave me his big hand.... In that somber room I felt the depth of undying friendship.

Maurice O'Sullivan, who grew up in the Blasket Islands, off the southwest coast of Ireland, in the early years of the 20th century, described how his sister reached the decision to emigrate.

Well," said Maria one day.... The rest of us were sitting around the fire. We turned.... "What is the 'well'?" asked my brother Shaun. She turned back to the table again, smiling. Then picking up a cup she turned around again. "'The well'...[is] that I have a great mind to go to America."

"What put that in your head?" said my father....

"Peg is going and I have no need to stay here when all the girls are parting."

"She won't go," said Eileen, her lips trembling, "or if she does, I will go too."

"Well, fly away at once!" cried my father, waving his hands in the air, "away with you over the sea and you will find the gold on the streets!"

There was nothing in Ireland. Thanks be to God I got away from it. There's nothing there. My mother's biggest mistake was that she never came over here. It was the biggest mistake she made in her life.

—Bridget Fitzgerald, who came to the United States in 1921

A man kneels to receive the blessing of a priest as the long journey from an Irish village to America begins.

27

LEAVING HOME

In 1848 Diarmuid O'Donovan Rossa had to say good-bye to his mother, brother, and sister, who had received ship tickets from a brother who had already gone to Philadelphia. Rossa would not be reunited with his family until 1863. He remembered the parting.

The day they were leaving Ireland, I went from Skibbereen to Renascreena to see them off. At Renascreena Cross we parted. There was a long stretch of straight even road from Tullig to Mauleyregan over a mile long. Renascreena Cross was about the middle of it. Five or six families were going away, and there were five or six carts to carry them and all they could carry with them, to the Cove of Cork. The cry of the weeping and wailing of that day rings in my ears still. [At] that time it was a cry heard every day at every Cross-road in Ireland. I stood at that Renascreena Cross till this cry of the emigrant party went beyond my hearing. Then, I kept walking backward toward Skibbereen, looking at them till they sank from my view over Mauleyregan hill.

A dockworker lowers the gangplank of a ship leaving Morville, Ireland, in 1900. Carrying her basket of food, an old woman starts down the steps to the steerage.

An observer described the farewells at a railroad station, where the emigrants were often accompanied by friends and family in a "convoy."

A deafening wail resounds as the station-bell gives the signal of starting. I have seen gray-haired peasants so clutch and cling to the departing child at this last moment that only the utmost force of three or four friends could tear them asunder. The porters have to use some violence before the train moves off, the crowd so presses against door and window. When at length it moves away, amidst a scene of passionate grief, hundreds run along the fields beside the [track] to catch yet another glimpse of the friends they shall see no more.

Many 19th-century Irish emigrants departed for the New World from the English port city of Liverpool. There they were given a brief medical examination before boarding a ship. An emigrant leaving in 1850 described his medical exam.

I passed before him [the doctor] for inspection. He said without drawing breath, "What's your name? Are you well? Hold out your tongue; all right," and then addressed himself to the next person.

At the port of departure, emigrants were often met by "runners" who took advantage of them by offering services they did not need. Frederick Sabel, a Liverpool rooming house owner, described how the runners preyed on the emigrants coming from Ireland to board ships bound for America.

T he runners watch the steamboats [bringing passengers from Ireland] night and day. Suppose a ship arrives at one o'clock in the morning, and it rains as fast as it can; the passengers are all frightened; everything is strange to them; and the moment the ship arrives, however many passengers there may be, there come as many runners, who snatch up their luggage as quickly as possible, and carry it away; they are like so many pirates. If any of the passengers wish to keep on board the ship, they cannot; the police come and drive them away.... Everything seems as if it were done on purpose to favour the runners and man-catchers; the people are actually driven from the ship with sticks into the hands of those people.... It is really a horrible thing.

In the 1870s, David S. Lawlor's father was a mechanic in a cloth mill in County Waterford, Ireland. He refused to testify against a fellow employee who was accused of stealing. The incident made him decide to immigrate to America. As his son recalled years later, the word itself had a magical sound.

A merica the land of promise! We all looked forward to the day we were to go. It was the dream of all the young people of Ireland to go to America. Every family tried to save enough to have one son or one daughter go, knowing that this child would work and save until enough money was sent home for another to have passage money.

Stephen Byrne, who in 1873 wrote a book of advice for Irish emigrants, counseled:

Young persons, then, having made up their minds to emigrate, ought to begin a religious preparation. If they are Catholics—and to these I principally address myself, because they are entirely the most numerous among English-speaking emigrants—they ought to receive worthily the sacraments of the church before leaving home. All Catholics know that this is a duty incumbent upon them in all undertakings involving danger, and also in entering upon a new condition of life. An unburdened conscience gives a man the freedom of heaven, and establishes him in peace with God and his fellow-men.

In this 1866 print, the driver blows his horn to announce the departure of a coach headed for a seaport. The emigrants are loading their baggage for the trip.

Make your bargain for your passage with the *owner* of the ship, or some well-known respectable broker or ship-master. Avoid by all means those crimps [con men, or hustlers] that are generally found about the docks and quays near where ships are taking in passengers. Be sure the ship is going to the port you contract for, as much deception has been practiced in this respect. It is important to select a well-known captain and a fast sailing ship, even at a higher rate.

The Irish Emigrant Society of New York was one of many organizations that offered help to the newcomers. This poster advertised the opportunity to send money home to pay for the ship passage of relatives in Ireland.

Irish Emigrant Society,

51 Chambers St., New York.

Incorporated April 29th, 1844.

REMITTANCES TO IRELAND.

Passage Tickets by all the Steamship Lines.

PERFECT SECURITY AND PROTECTION.

BILLS AT SIGHT ON THE BANK OF IRELAND and all its BRANCHES, for One Pound and upwards, sold at the lowest market rates.

STEERAGE PASSAGE TICKETS to or from Ireland can be procured at the Office at the lowest price of the several lines of steamers sailing to and from New York.

Persons residing out of the city, who desire to remit money (or passage tickets) to the Old Country through the Society, should carefully state to whom and where it is to be sent, with name of *Town, Parish, Post-Office,* and *County.*

☞ All communications should be addressed to W. P. BYRNE, Treasurer of the IRISH EMIGRANT SOCIETY, No. 51 Chambers Street, New York, and all Checks and Drafts made payable to the order of the IRISH EMIGRANT SOCIETY.

THE OBJECTS OF THE SOCIETY ARE

1st. By the sale of its bills on the BANK OF IRELAND, and by keeping ample funds there to meet the same, to furnish the Irish people in this country a safe remittance to their friends in Ireland, thereby protecting them from false, fradulent, and impious bills.

2d. To receive money from persons having no knowledge of responsible shipping houses, and to purchase for them passage tickets by steamers or sailing vessels, and to transmit the same to their friends in Ireland, for whom the passages were ordered.

3d. To receive moneys for persons residing out of the State of New York, and who have no responsible agent here, and to apply such moneys to forwarding into the interior, according to their directions, their friends and families who may arrive at this port from Ireland.

4th. By and through the official position of the President (he being by law of the State of New York a Commissioner of Emigration), to protect the persons and property of Irish Emigrants.

5th. To aid friendless Irish Emigrants and their children, by an application of the surplus earnings of the Society to their relief.

6th. To afford advice, information, and protection to Emigrants from Ireland, and generally to promote their welfare.

JAMES LYNCH, President.
JOHN H. POWER,
JAMES OLWELL, } Vice-Presidents.
JEREMIAH DEVLIN,
JEREMIAH J. CAMPION, Recording Secretary.
DANIEL O'CONOR, Corresponding Secretary.

MEMBERS OF THE EXECUTIVE COMMITTEE.

James Olwell,	Jere. J. Campion,	Jere. Devlin,
Bryan Lawrence,	Edward Bayer,	Lewis J. White,
Edward Frith,	Euwene Kelly,	James Lynch,
Henry L. Hoguet,	John H. Power,	R. O'Gorman,
W. Von Sachs,	H. J. Anderson,	J. B. Nicholson,
James Wallace,	E. C. Donnelly,	T. H. O'Connor,	Daniel O'Conor.

SALARIED OFFICERS.

It was planned that my father should come first and find a home for us. I will never forget the day he left home, because mother baked a loaf of currant bread for him to take with him. She offered it to him at the station as the train was leaving and he refused to take it. I could have killed him I was so angry. Mother told me never to be angry, that it was a grievous sin, so never since that day have I lost my temper.

[David Lawlor's father found work in a mill in Massachusetts and soon sent for his family.] The night before we were going away the neighbors gave us a dance. Nearly the whole town was there sometime during the evening. I know I went to sleep on the settle bed while the fiddles were playing and the feet were dancing.

In the morning we took the train for Cork and the young and the old were at the station to wish the Lawlors "Goodby, good luck and God bless you." The Irish cry easily, they cry for joy and for sorrow; maybe that is what keeps their hearts so soft and so warm.

Encouraged by his brother Frank, who had immigrated to New York City, Paul O'Dwyer decided to leave his home in County Mayo in 1925. He recalled his leave-taking.

There was a custom, which must have grown up in the famine of 1848, that was known as the "American wake." It occurred on the eve of an Irish emigrant's departure for the United States. In those days most emigrants never returned, hence the term "wake." My relatives and neighbors gathered in the house, stood around and encouraged me. They said such things as, "Well, you're going to be with your brothers, so it will be just like home." I knew that was not true, but I smiled just the same. The older people were saddened, and I had mixed emotions. I feared going to America, but I knew there was nothing for me in Mayo....

The neighbors left about midnight. Each one pressed a coin into my hand. The sum came to seven dollars in all, a tremendous amount for the poor of our parish to part with. My mother had purchased a new suit for me, tightly fitted and in

Emigrants sometimes paid for their ship tickets at an emigration agency, as in this newspaper illustration from 1856.

keeping with the latest Irish style. It was a blue serge suit and the bottom of the jacket barely came to my hips.

The next morning my mother and sisters accompanied me on the trip to the railroad station by pony and trap [cart]. There were periods of silence when we faltered in making the best of it. At the station my sisters cried, and my mother didn't. It wasn't manly to cry, so I didn't either—until the train left the station. Then I did. I felt bereft and terrified.

A woman remembered her own American wake when she left Ireland in the 1890s.

It would not have been so bad. Only in the morning every one said so-long to you and you would know by them that they never expected to see you again. It was as if you were going out to be buried.

Kitty Fitzgerald Donovan came to the United States when she was 19 years old. One of 12 children, she grew up in Newcastle West, County Limerick. Her older brother was already in Chicago.

I left in 1927. I went to the port of Cobh, that used to be called Queenstown, where you were tested for health and mentality. The test was not very rigorous, you know they would stop you if you had something like TB [tuberculosis]. Then you got your sailing date if you passed the test.

This poster from the 1880s shows one of the ships of the White Star Line, which took many Irish immigrants across the Atlantic. The regular route started at Liverpool, England, and stopped the next day at Queenstown, in Ireland.

The Irish port of Queenstown (now called Cobh), as an artist depicted it in 1874. The trunk in the lower right-hand corner is marked "M. Fitzgerald, Passenger, New York, Steerage."

Joseph Patrick Fitzpatrick, who came from Ireland to the United States in 1920, recalled:

I had a small steamer trunk for a start. One of those small ones that you can push underneath a bunk. I didn't bring very much clothes, just a work suit and my best suit. And I had this pound of butter wrapped up. I guess somebody told me to take it to somebody who'd like Irish butter. It was good butter my sisters made.

Bunks in the steerage section of a ship provided little room for the passengers. In the 1870s, the journey from Ireland to America took only a week or two in a steamship. The sailing ships that brought the famine refugees 20 years before could take up to two months to cross the Atlantic.

CROSSING THE ATLANTIC

Thomas Cather kept a journal of his trip to America in February 1836. He traveled with a friend, Henry Tyler; both were comparatively wealthy young men who decided to make the trip on a whim. Though he eventually returned to Ireland, Cather provides in his journal a vivid portrayal of what an emigrant's journey aboard a 19th-century sailing ship could be like.

26th February—

Yesterday evening the wind freshened, then came on several squalls at short intervals and then it fairly set in to blow a gale. Awful tumbling during the night. This morning it still blows great guns.... The sea presents one of the most magnificent sights I have ever witnessed. As far as the eye can reach is a confused, tortured mass of roaring surges, the ridges of the waves are all roughened and broken into foam, while here and there an immense black billow, like a mountain in motion, comes sweeping on high over the other surgest till, culminating in a lofty top, it tosses up a snow-white cloud of spray, with a loud explosion. It was worth while coming so far to see the ocean in an angry mood. It has just struck eight bells [twelve o'clock]. Henry and I are the only passengers able to be up. The rest are in their berths groaning in agony of spirit. It is with difficulty I can write. Every movable in the cabin seems to be instinct with life, or rather possessed by a devil. It was with no small trouble we managed to get breakfast over. During that repast a pot of butter launched itself at Henry's head with the velocity of an arrow; a large piece of corned beef jumped off the table and bolted through the open door...and a dish of pork chops bounded off the board and set to jigging it on the floor very merrily. However, we managed to get hold of some eatables before they all eloped....

27th February—Last night it blew tremendously hard. Today the gale has increased. We are almost under bare poles. [Most of the sails have been furled.] A ship on the weather quarter is lying to, about a mile distant. We just catch a view of her as she rises on the crest of a wave. The sea presents a most sublime spectacle—a howling wilderness of foaming waters. Two of the sails have been blown to rags.

Few emigrants traveled in such style. Another mid-19th-century passenger described the rough treatment he and others experienced while boarding a ship leaving Liverpool.

Men and women were pulled in any side or end foremost, like so many bundles. I was getting myself in as quickly and dextrously as I could, when I was laid hold of by the legs and pulled in, falling head foremost down upon the deck, and the next man was pulled down upon the top of me. I was some minutes before I recovered my hat, which was crushed as flat as a pancake. The porters, in their treatment of the passengers, naturally look only to getting as much money as they possibly can from them in the shortest space of time, and heap upon them all kinds of filthy and blasphemous abuse, there being no police regulations, and the officers of the ship taking the lead in the ill-treatment of the passengers.

In 1975 Richard Howard Brown set down the story his mother had told him of the ocean voyage her grandfather made to America in 1848. In that famine year, many were already suffering from disease when they boarded the ships. Richard Brown's great-grandfather was a five-year-old boy who made the trip with his uncle.

The saving and preparation for their trip took almost a year because few of the ship owners trafficking in emigrants provided food or sleeping quarters. Transporting the Irish to the New World in the time of the Famine was a speculative business, which meant that owners invested as little as possible and moved as many as they could crowd onto a ship. Travelers provided their own bedding and provisions and payment for the trip covered passage only and a supply of water, which on the ship my great-grandfather sailed on was down to a cup a day per person within a month because most of the casks leaked and two had previously contained vinegar and the contents were undrinkable.

There were only two stoves on deck and the cooking lines were long. When winter storms kept everyone down below, the food was eaten raw. The supplies of some of the passengers ran out early in the voyage, either through miscalculation or because portions that had been gnawed by rats had been thrown overboard to feed the fish. Some families shared what they had with others but still there were fights for the food that remained and two men were killed, and the fastidiousness of the first weeks was forgotten when some of those same people who had thrown food away tried to catch the rats so they could eat them.

The ship landed...one day short of seven weeks after leaving Sligo Bay. Most of those who disembarked were weak and sickly, a sorrier lot by far than when they started. Of the four hundred who set out from Ireland with my great-grandfather, sixteen never survived to see the fabled New World that was going to change their lives. Except for the few killed in fights for food, they died of dysentery and typhus and were thrown overboard as soon as they were dead.

New York politician Paul O'Dwyer recalled a song that the emigrants sang aboard ship, titled "I'm Off to Philadelphia":

When they told me I must leave the place
I tried to keep a cheerful face,
To show my heart's deep sorrow I was scorning,
But the tears will surely blind me,
For the friends I lave behind me,
When I'm off to Philadelphia in the mornin'.

With me bundle on me shoulder,
Sure there's no one can be bolder,
And I'm lavin' dear old Ireland in the mornin'.
For I lately took a notion,
For to cross the Briny Ocean,
And I'm off to Philadelphia in the mornin'.

Above and below decks, this sailing ship was crammed with passengers in 1840. Conditions became even worse later in that decade when the famine drove hundreds of thousands of people out of the country.

Passengers who could afford to travel first-class enjoy "a spot of beef tea" in deck chairs on the ship Ethiopia in 1900.

On the same voyage as the one above, a group of Irish children from the steerage section sit on the deck. The voyage was hard for children, who hated the strange food and had nothing to occupy their time.

Ann McNabb described her trip in the mid-19th century.

I sailed from Londonderry. The ship was a sailin' vessel, the "Mary Jane." The passage was $12. You brought your own eating, your tea an' meal, an' most had flitch [bacon]. There was two big stoves that we cooked on. The steerage was a dirty place and we were eight weeks on the voyage—over time three weeks. The food ran scarce, I tell you, but the captain gave some to us, and them that had plenty was kind to the others. I've heard bad stories of things that went on in the steerage in them old times—smallpox and fevers and starvation and worse. But I saw nothing of them in my ship. The folks were decent and the captain was kind.

Twenty-eight-year-old Henry Johnson crossed the ocean in the summer of 1848. After an eight-week voyage, he arrived safely in New York. But as he wrote his wife, Jane, who remained with their children in County Antrim, the food he had taken on board was so bad he nearly starved.

The pigs wouldn't eat the biscuit so that for the remainder of the passage I got a right good starving. There was not a soul on board I knew or I might have got a little assistance, but it was every man for himself....

[Johnson's ship encountered a storm.] Anything I have read or imagined of a storm at sea was nothing to this.... One poor family in the next berth to me whose father had been ill at the time...I thought great pity of. He died the first night of the storm and was laid outside of his berth. The ship began to roll and pitch dreadfully. After a while the boxes, barrels &c began to roll from one side to the other, the men at the helm were thrown from the wheel, and the ship became almost unmanageable. At this time I was pitched right into the corpse, the poor mother and two daughters were thrown on top of us and the corpse, boxes, barrels, women and children all in one mess were knocked from side to side for about fifteen minutes. Pleasant that wasn't it. Jane Dear Shortly after the ship got righted and the captain came down we sewed the body up [in a shroud], took it on deck and amid the ragin of the storm he read the funeral service for the dead and pitched him overboard.

Stephen de Vere came to the United States during the Great Famine. He described conditions in the steerage of a ship.

Before the emigrant has been at sea a week he is an altered man. Hundreds of poor people, men, women, and children of all ages from the drivelling idiot of 90 to the babe just born, huddled together without light, without air, wallowing in filth and breathing foetid atmosphere, sick in body, dispirited in heart...the fevered patients lying between the sound in sleeping places so narrow as almost to deny them...a change of position...by their agonized ravings disturbing those around them...living without food or medicine except as administered by the hand of casual charity, dying without spiritual consolation and buried in the deep without the rites of the church!...

The food was seldom sufficiently cooked because there were not enough opportunities for drinking and cooking. Washing was impossible; and in many ships the filthy beds were never brought up on deck and aired, nor was the narrow space between the sleeping berths washed or scraped until arrival at quarantine. Provisions, doled out by ounces, consisted of meal of the worst quality and salt meat; water was so short that the passengers threw their salt provisions overboard—they could not eat them and satisfy their raging thirst afterward. People lay there for days on end in their dark close berths, because by that method they suffered less from hunger. The captain used a false measure of water, and the so-called gallon measure held only three pints.... Spirits were sold once or twice a week, and frightful scenes of drunkenness followed. Lights below were prohibited, because the ship, in spite of the open cooking-fires on her decks, was carrying a cargo of gunpowder.... The voyage took three months, and...many of the passengers became utterly debased and corrupted.

David S. Lawlor was seven when he traveled to America with his mother, brothers, and sisters. He recalled:

The voyage lasted nine days and I had a happy and hungry time. I was always hungry, all the other boys being hungry too. I remember my mother giving her two potatoes to a couple of boys she had tucked under her wing.

We came into New York harbor one glorious morning the fourth of July, 1872. It was Sunday and they would not let us land. We could see the people going to mass so we united our prayers with theirs and thanked God for bringing us safely across the deep. This good ship on her next passage was lost at sea with all aboard.

Kitty Fitzgerald Donovan remembered her trip happily. She emigrated from County Limerick as a young woman of 19 in 1927. In the 20th century shipboard conditions for the emigrants were vastly improved:

My trip took seven days in steerage. It was fun but hard too. Sometimes the motion of the boat could make some sick. I had no trouble luckily. And there was lots of food and enjoyment. They had games for the travelers. It was like a week's vacation really, if you could stand the ship's movement. The food was good.

After land was sighted, everyone came on deck, hopeful and eager to disembark onto their new country.

The Carrolls of Maryland

The first Carroll in America was an Irish aristocrat who was granted lands in Maryland by King James II. Though the colony had been founded as a refuge for Catholics, it soon had a Protestant majority, and laws were passed to restrict Catholics' rights. They were prohibited from voting and holding office and forbidden to worship publicly. The Carroll family, though remaining faithful to Catholicism, held onto most of their land and wealth. Two of its members were prominent in the independence struggle that formed a new nation, the United States.

John Carroll, born in 1735, went to France as a young man, where he was ordained as a priest. Returning to America in 1774, he joined the English colonies' struggle for independence. In 1789, after the fight had been won, he founded a school that became Georgetown University, the first Catholic college in the United States. The following year, he was consecrated as the first American Catholic bishop and in 1808 became archbishop of Baltimore. At his death in 1815, John Carroll left behind a strong American Catholic church.

His cousin Charles was born in 1737 in Annapolis, Maryland. Educated in the law in France and England, he returned to America in 1765, and his father turned over to him the estate, called Carrollton. Charles Carroll threw his wealth behind the independence movement. With his cousin John he went to Canada to persuade French Catholics to support the fight for independence. In 1776 he was chosen as a Maryland delegate to the Continental Congress, where he was the only Catholic to sign the Declaration of Independence.

Charles Carroll served as a U.S. senator in the first Congress. Later he devoted his talents to developing the new nation's transportation system. He built canals and was chairman of the Baltimore & Ohio, the first railroad in the United States. When he died in 1832, he was the last surviving signer of the Declaration of Independence and one of the richest men in America.

THE IRISH IN COLONIAL AMERICA

Many of the Irish who came to the English colonies in America arrived as indentured servants or redemptioners. The practice led to abuses, and a committee of the House of Commons began an investigation of the treatment of indentured servants. The Belfast News Letter *of March 22, 1774, described the testimony of Dr. Williams, an American colonist.*

It appeared that a trade was carried on in human flesh between Pennsylvania and the province of Ulster. Such of the unhappy natives of that part of Ireland as cannot find employment at home, sell themselves to the masters of vessels, or persons coming from America to deal in that species of merchandise. When they are brought to Philadelphia...they are either sold aboard the vessel, or by public vendue [auction], which sale on arrival there is public notice given of, either by handbill, or in the newspapers. They bring generally about fifteen pounds currency at market, are sold for the term of the indentures, which is from two to four years, and on its expiration, receive a suit of clothes, and implements of husbandry, consisting of a hoe, an axe, and a bill from their taskmasters. Several gentlemen in the committee expressed their abhorrence of such a barbarous traffic.

Sometimes indentured servants ran away from their masters before their time of servitude was over. The following notice appeared in the Pennsylvania Gazette *on May 19, 1751.*

Run away from Thomas James, of Upper Merion, Philadelphia County, on the 5th of this inst., an Irish servant lad named William Dobbin, about eighteen years of age, speaks good English, fresh colour'd, thick and well set in his body, has light colour'd curled hair, somewhat resembling a wig. Had on when he went away an old felt hat, ozenbrigs [coarse linen] shirt, and old dark born colour'd coat, too big for him, and breeches of the same, grey worsted stockings, and a pair of old shoes, with brass buckles, one of the buckles broke. Whoever takes up and seizes this servant so that his master may have him again, shall have twenty shillings reward, and resonable charges, paid by Thomas James.

It was possible for indentured servants to prosper after serving out their required term of employment. Other Irish immigrants also found success, as indicated in the following letter from James Murray, an immigrant from Ireland living in New York City, to the Reverend Baptist Boyd of Aughelow in County Tyrone, on November 7, 1737.

Read this letter, and look, and tell aw the poor Folk of your Place, that God has opened a Door for their Deliverance; for here is ne scant of Bread, and if your Sons Samuel and James Body wad but come here, they wad get more Money in ane year for teechin a Lettin Skulle, nor ye yer sell wad get for Three Years Preechin whar ye are.... The Young Foke in Ereland are aw but a Pack of Couards, for I will tell ye in short, this is a bonny Country, and aw things grows here that ever I did see grow in Ereland; and wee hea Cows and Sheep and Horses plenty here, and Goats, and deers, and Raccoons, and Moles, and Bevers, and Fish, and Fouls of aw Sort.... There is Servants come here out of Ereland, and have serv'd their Time here, wha are now Justices of the Piece.

Even in colonial times, letters sent back to Ireland attracted more immigrants. Here is part of a letter from a resident of Monmouth County, New Jersey, written on Sunday, March 20, 1796.

Dear Mother I want you to come here very much and live with Me and you can live better than in Best Mans house in Ireland.... Dear Brother Come here without fail as you Can work at your trade.... You can get a shilling a yard for weaving anything.... Bring your loom.... Dear Brother if you have any spare money bring me a good riding saddle as they are dear here.... Likewise Coper tea kettles is very Dear here. Buy two or three and use them once and you wont Pay any Duty for them.

Irish Servants.

JUST ARRIVED, *in the* Ship JOHN, *Capt.* ROACH, *from* DUBLIN,
A NUMBER of HEALTHY, INDENTED
MEN and WOMEN SERVANTS:
AMONG THE FORMER ARE,
A Variety of TRADESMEN, with some good FARMERS, and stout LABOURERS: Their Indentures will be disposed of, on reasonable Terms, for CASH, by
GEORGE SALMON.
Baltimore, May 24, 1792.

Many of the Irish immigrants in colonial times were indentured servants, obligated to serve a master until the cost of their passage was paid. Like slaves, indentured servants could be sold at auction and sometimes escaped from those who bought their services.

John Barry

In April 1776 Captain John Barry took the British ship *Edward,* the first American prize of the revolutionary war, into Philadelphia harbor. This was only the first of a string of victories that would earn John Barry the title "father of the American navy."

John Barry was born around 1745 in County Wexford, Ireland. The boy absorbed the county's seafaring tradition and in his teens shipped off to the American colonies. He settled in Philadelphia and became a merchant shipmaster. When the American Revolution broke out, Barry outfitted his own ships to support the Patriots.

He brought supplies to the Continental Army and raided British supply ships. In 1780 Barry was given command of the *Alliance* to carry an American diplomat to France. On his return, as the *Alliance* lay becalmed in the waters off Newfoundland, two British ships attacked. Though Barry was wounded in the fight, he managed to turn the tables on the enemy when a breeze reactivated his badly mauled ship. Bringing his guns to bear, he captured both of the British ships.

In the last years of the war, he continued his unbroken string of victories over British ships. The *Alliance* successfully fought the last naval battle of the war in the waters around Florida just before the peace treaty was signed in 1783.

Barry went back to his merchant shipping business. However, he was recalled to service to lead the new nation's navy in 1794. In that role he commissioned new battleships and trained many young officers who would serve the nation well. At his death in 1803, Americans mourned for "the father of the American navy."

A group of hopeful Irish immigrants arrives in New York City in 1900.

CHAPTER THREE

PORTS OF ENTRY

After the American Revolution, the United States provided a haven for Irish fleeing economic and political repression. Between 1783 and 1820, about 150,000 people left Ireland for the United States. A sizable number of these newcomers were middle-class Protestants (Presbyterians) from Ulster.

The proportion of Catholics among the Irish newcomers began to rise during the 1820s. Many immigrants landed in New York City, which by now was the chief port of the young United States. The construction of the Erie Canal (1818–25) in upstate New York gave immigrant laborers an opportunity for work. But coastal cities such as Philadelphia, Baltimore, and Boston still drew significant numbers of immigrants in the prefamine years. Other immigrants sailed directly to New Orleans, the gateway to the rich farmlands of the Mississippi Valley.

Even so, by 1845—the first year of the famine—three out of four immigrants still arrived in New York. In that year, one-third of the population of the nation's largest city had been born in Europe, most in Ireland or Germany.

Stepping off the boat, the immigrants faced a crowd of "runners" who were ready to bilk them. Some runners worked for boarding-houses that charged high prices to inexperienced newcomers. Other swindlers offered immigrants over-priced or phony railroad tickets, took their money in return for the promise of a job, or engaged in other kinds of scams that were designed to separate the greenhorns from their savings. Many immigrants thus cheated were reduced to begging or finding a bed in one of New York's almshouses, shelters for the homeless.

In 1847 as Irish refugees from the Great Famine flocked to the United States, the New York State legislature set up a commission to find ways of dealing with their various problems. The commission established a hospital to treat those who suffered from diseases or malnutrition and also drew up regulations designed to keep unwary immigrants from being exploited.

In 1855 the New York emigration commissioners opened a landing depot at Castle Garden, a former theater on an island just off the southern tip of Manhattan Island. For the next 35 years, 70 percent of all immigrants took their first steps on American soil at Castle Garden. There they received information about jobs and lodging, found secretaries to write letters home for them, and exchanged foreign currency at fair rates. Castle Garden was the chief port of entry during the peak years of Irish immigration, welcoming more than 2 million Irish during the years of its operation.

By 1890 great numbers of immigrants were arriving from other nations as well, and Castle Garden was no longer adequate to process them all. The federal government took over the job, opening a larger immigrant landing station at Ellis Island in New York Harbor. The very first immigrant to go through Ellis Island, on January 1, 1892, was a 15-year-old Irish girl from County Cork named Annie Moore. Another 1 million Irish immigrants followed her until Ellis Island was closed in 1954.

Today's newcomers arrive on jumbo jets, landing in New York, Boston, Chicago, and other cities that have large Irish American communities. Many arrive with tourist or student visas but intend to find jobs. Technically illegal aliens, these immigrants follow the path that their predecessors have blazed for four centuries.

ARRIVAL

This painting shows Irish immigrants landing at the Battery, the southern tip of Manhattan Island in New York City. Before 1855, they simply stepped ashore and began the adventure that they hoped would lead to success.

The Irish immigrants' first experiences in an unfamiliar country could be quite terrifying. Rosalie B. Hart Priour arrived in 1834, when she was about 10. She and her family, who were headed for Texas, landed at New Orleans, where even the dockworkers were a strange new sight.

None of us had ever seen a negro before, and the children were nearly frightened to death. They were a great curiosity, even for the grown people. There was very little to make us feel satisfied. We arrived at a time when the cholera was raging in New Orleans; people were dying so fast that it was impossible to dig graves, and the dead were buried in trenches. The immigrants were forbidden to eat vegetables or fruit, as it was supposed that they were dangerous. The majority obeyed orders, but others disregarded the rules...and were the only ones who escaped.

In the 1840s an Irish immigrant in New York City found himself set upon by two "runners," people who drummed up customers for boardinghouses. A reporter summarized his story.

The moment he landed, his luggage was pounced upon by two runners, one seizing the box of tools, the other confiscating the clothes. The future American citizen assured his obliging friends that he was quite capable of carrying his own luggage. But no, they should relieve him...of that trouble. Each [represented] a different boarding house, and each insisted that the young Irishman...should go with him.... Not being able to oblige both gentlemen, he could oblige only one; and as the tools were more valuable than the clothes, he followed in the path of the gentleman who had secured that portion of the "plunder."

In 1855, Castle Garden opened as an immigrant landing station in New York and offered help and advice to the newcomers. This drawing shows both Irish and German arrivals, the two major immigrant groups of the mid-19th century.

Partly to protect the immigrants from such runners, New York State opened an immigrant receiving station at Castle Garden in 1855. Father Stephen Byrne, who wrote a book of advice for would-be immigrants, described conditions at New York's Castle Garden in the 1870s.

I t is essentially an institution of protection to the immigrant. It is as truly a work of mercy as a hospital or an orphan asylum. Not that the funds for its support come from the State or private charity; on the contrary, they are contributed by the immigrant himself, by a tax of one dollar and fifty cents on every one that arrives, which is paid out of his passage-money....

All immigrants are obliged to land at Castle Garden, where they are provided with temporary accommodations suitable to their requirements. Those who have tickets for the interior, or money to take them to any point outside of New York, are immediately put upon one or other of the great railroad lines diverging from the city to all parts of the country, without any trouble or risk to the immigrant.... Any immigrants having gold, silver, or uncurrent money of any kind can have it changed into the current money of the United States, also, at Castle Garden.

Father John Joseph Riordan, a priest from an Irish American organization that aided immigrants, had an office at Castle Garden. In 1886 he wrote:

I was not long at Castle Garden before it became apparent that there was a great work to be done. Every other day brought its shiploads of immigrants, who, after they passed through the hands of registration clerks, took their places in the Labor Bureau to wait for employment. Where were they to go at night, if an employer did not turn up in the meanwhile? Their only alternative hitherto had been to go indiscriminately with the first lodge-house keeper who got possession of them. For any one acquainted with the life of a great city it is unnecessary to dwell on the dangers to which virtuous young girls and unsophisticated young men were thus exposed.

In 1892 a new immigration station opened in New York Harbor on Ellis Island. Newcomers now had to pass through a series of physical and mental examinations. Some were rejected because they were mentally or physically incapacitated and therefore likely to become wards of the state. Elizabeth Phillips came from Ireland in 1920 when she was 19; 66 years later, she recalled the sometimes frightening experience:

T his boat took us straight to Ellis Island and I remember long counters, I'm looking in my mind I can see them.... I remember we come to two people, and they looked at your passport and asked you questions and confirmed it to that other person and from there we went to the next two people and they examined, they pulled the eyes down, they looked in our ears, our mouth and I believe they listened to our chest. I'm

Castle Garden, where immigrants entering New York landed in the 19th century, was connected to Manhattan Island by a bridge. Relatives and friends often waited for the newcomers to arrive. This reception was described in the New York Sun *in 1880:*

One day last week a young Irish girl, who had come over alone, was sighted in the passageway between the rotunda and the Information Bureau...by four women relatives who had gone down to meet her. Her back was toward them, and she was unaware of the proximity, until with a "whillillilew" they precipitated themselves on her.

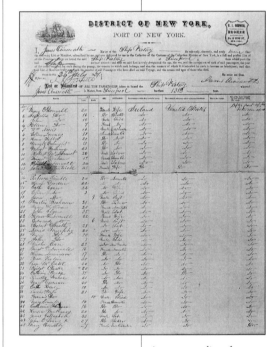

A passenger list of a ship arriving in New York from Liverpool in 1861 shows that most new arrivals were in their early 20s or younger. They were among the ambitious immigrants who enriched the United States with their talents and labor. (The word Do *in this list meant "ditto," or "same as the word above.")*

One of the frightening experiences at Ellis Island was an examination of the eyes. The inspectors were looking for a contagious disease called trachoma. Two of the women have the letter E marked in chalk on their coats. It indicates that another inspector had suspected them of having eye trouble.

pretty sure and I remember one of them saying, "Why are you frowning for?" I guess I must have been frowning because [I had noticed some other people nearby].... To me they were foreign, people you know. There were a lot of children and they were all dressed in heavy clothes, they didn't speak English... and they had little tiny shawls or babushkas or what do you call them...and some of them standing there crying.... I heard them cry, cry, and I asked somebody why and they said, "Because they're rejected."

Emanuel Steen, born in Dublin in 1906, came to America in 1925. Interviewed in 1991, he still resented the humiliating way he had been treated at Ellis Island.

Everybody was tagged. They didn't ask you whether you spoke English or not. Everybody was tagged. They took your papers and they tagged you. That was the first thing. They checked your bag. They had to go through your baggage and then they pushed you, you know what I mean? They just pushed you. They'd point because they didn't know whether you spoke English or not. They could have looked [on the tag] to see, I suppose, but they didn't. Nobody asked me. They had too many. Understaffed. Overcrowded. Jammed. And the place was the noisiest and the languages and the smell....

INTERVIEWER: *What did it smell like?*

STEEN: Foul, you know what I mean? But I am nineteen [at that time]. I can stand a lot then...and I just, it didn't bother me to that extent. You know, you figure, "Get out of here fast," you follow me? Then you had to go through the physical. I think, frankly, the worst memory I have of Ellis Island was the physical because the doctors were seated at a long table with a basin full of potassium chloride and you had to stand in front of them, follow me?.... And you had to, uh, reveal yourself.... Right there in front of everyone, I mean, it wasn't private! You were standing there. And the women had to open their blouse and here this is terrible. Remember, these were immigrants from a very reticent people. And here, nobody was looking or watching anyone. Looking back I can see that but I was nineteen and I was embarrassed as hell, you know?.... The place was jammed, follow me? And again, we're not aware this is historic and this is something you're going to tell your grandchildren about, that I would be talking to you someday about it. I just want to get through there and get out.

Aileen Vroom was born in 1905 in Waterford, Ireland. She came through Ellis Island at the age of 24. Years later, she indignantly remembered the examination of her hair for lice.

The woman who examined my hair wasn't a friendly type. She took a fine comb and started to comb my tangly head. The tears were running out of my eyes because of the pull. She didn't find anything, so I got by.

But the girl I had come over with...she wasn't so lucky.... Poor Mollie passed, but she had to have a sulphur bath. And then on her suitcase, they put a big label: "Inspected, Disinfected and Passed." She smelled like a terrible sewer. Right after we got out, I took her to have her hair washed...and I tore the label off her suitcase, whether it was right or wrong.... I didn't see why she should carry that banner around with her.

Today the port of entry is more likely to be a major airport. Ann, an illegal immigrant from Ireland who lives in New York City, described how her friend, who did not have a U.S. work permit, was turned back.

She was coming in from Ireland to work for a family. Everything had been arranged through friends—she even carried a letter in her bag from the family who was hiring her, giving all the details of the job. She must have felt guilty and showed it or maybe Immigration questions thoroughly every tenth person or something. Anyway, they started in on her and she broke down and cried. Even showed the letter [from the family she was going to work for]. Well, she was on the next plane back to Ireland. It was too bad—for her and for the family because they were counting on her.

Ann herself has learned "the secret" of getting past the immigration officials. She was even able to reenter the country after going to Ireland for a while.

When I went back to Ireland for a visit, I really was afraid I wouldn't get back in to the States, so I made sure I had my return ticket and plenty of money for a two-week stay. That's all I asked for [when applying for a visa]—just two weeks. That way, they don't get suspicious.

What makes all this possible is the passport system. They stamp when you enter but not when you leave. So all my passport showed was that I came to New York once in 1972. Didn't say I left in 1976. If they asked, I was prepared to say I stayed only two weeks in 1972 but they never even asked.

The secret really is in acting like what you're doing is all right. I remember I went right into a bank and opened an account. Nobody told me I couldn't and I had a little money with me. Opened a savings account. When they asked me for my social security number, I just said, 'Oh I don't have that yet,' and they didn't blink an eye. Just said to bring it in when I got it. Same thing when I went to get a driver's license. Took the test like everybody else and got a license. Nobody asked me to prove where I was born. Course if you didn't speak English, you might have a bit more trouble. Or if you're black, that's probably very different.

The very first immigrant to pass through Ellis Island when it opened in 1892 was a teenaged Irish girl. The New York Times *reported the scene:*

The new buildings on Ellis Island constructed for the use of the Immigration Bureau were yesterday formally occupied by the officials of that department.... There were three big steamships in the harbor waiting to land their passengers, and there was much anxiety among the new-comers to be the first landed at the new station. The honor was reserved for a little rosy-cheeked Irish girl. She was Annie Moore, fifteen years of age, lately a resident of County Cork, and yesterday one of the 145 steerage passengers landed from the Gion steamship *Nevada.* Her name is now distinguished by being the first registered in the book of the new landing bureau.

The steamship that brought Annie Moore arrived late Thursday night. Early yesterday morning the passengers of that vessel were placed on board the immigrant transfer boat *John E. Moore.* The craft was gayly decorated with bunting and ranged alongside the wharf on Ellis Island amid a clang of bells and din of shrieking whistles.

As soon as the gangplank was run ashore, Annie tripped across it and was hurried into the big building that almost covers the entire island. By a prearranged plan she was escorted to a registry desk....

When the little voyager had been registered Col. Weber presented her with a ten-dollar gold piece and made a short address of congratulation and welcome. It was the first United States coin she had ever seen and the largest sum of money she had ever possessed. She says she will never part with it, but will always keep it as a pleasant memento of the occasion. She was accompanied by her two younger brothers. The trio came to join their parents, who live at 32 Monroe Street, this city.

Kitty Fitzgerald Donovan recalled her arrival from County Limerick in 1927 at the age of 19:

I landed in New York. Then the Traveller's Aid took over and put me on the train to Chicago. My brother met me there.... About a week later I started work at Sears Roebuck and stayed there for 15 years until I got married. My first impressions of the United States was that the country was so large. The scale was much larger than what I was used to.

FIRST IMPRESSIONS

Thomas Cather, a young Irish traveler who arrived in New York in 1836, looked up some of his friends who had immigrated earlier. In his journal, Cather commented that they were "terribly yankeefied." Just what that term meant can be guessed from his first impressions of American life.

8th March—

The Americans appear to be essentially a locomotive people. They are continually on the move.... There is always a crowd in motion to and from the principal marts.... At this hotel there are 150 or 200 persons, who all mess [eat] together.... The Americans are great economists of their time, and in nothing do they show their dispatch more than in eating. The first morning at breakfast I was absolutely astounded at the rapidity of their jaws. Hominy, Indian cakes and treacle, broiled fish, beef steaks, mutton chops etc. in "the twinkling of a bed post" were ravenously consumed; and before I had half finished my first cup of coffee the tables were cleared and I was left alone, like the last rose of summer. But the feats at dinner were still more astonishing.

During the Civil War, army recruiting agents met Irish and German immigrants at Castle Garden, offering them a cash bounty to sign up for military service.

Many Irish immigrants were impressed by the attitudes toward class differences in the United States. Around 1858 a Philadelphia laborer reflected:

People that cuts a great dash at home when they come here they think it strange for the humble Class of people to get as much respect as themselves, for when they come here it wont do to say I had such and was such and such back at home.... Strangers here...must gain respect by their conduct and not by their tongue. I know people here from the town of Newbridge that would not speak to me if they met me on the public road and here I can laugh in their face when I see them.

Sometimes the newcomers were disappointed by what they saw in America. In 1850 Joseph Brennan expressed his shock at the condition of the Irish in the United States in a letter that he mailed to Dublin.

It is time to speak of ourselves. I have been much disheartened since my arrival here by the unfortunate condition of my countrymen. I came with high hopes and sanguine expectations, and I have realized my disappointment.... Religious bigotry and party feuds have crossed the Atlantic with our people. Our nature has not changed with the clime.... What I state of the Irish in America is fact, and it is foolish or criminal to conceal it. Their position is not what it is represented to be at home—far different; it is one of shame and poverty. They

Josephine Cassidy, who went to Newark, New Jersey, in 1929 at age 13, remembered:

All I saw were tall buildings and tenement houses. I had never seen them before. I was like in another world. Because when you're coming from Ireland at thirteen you're really like seven. You're not as advanced as the kids in this country.

In 1896, a newly arrived immigrant poses with his baggage at South Ferry, at the southern end of Manhattan. The number of Irish-born Americans reached a peak of 1.8 million in the census of 1890 but declined steadily thereafter.

An immigrant boardinghouse was usually the first stop for new arrivals in the 19th century. The crowd here is sharing a meal, resting from the journey, and examining a set of knives offered by a salesman.

are shunned and despised. The name of Irish politics is anathema, and Ireland is as much a subject of contempt as pity. "My master is a great tyrant," said a negro lately, "he treats me as badly as if I was a common Irishman."

Playwright William Alfred related a story that his great-grandmother had told him of her arrival in the United States around 1866, when the Civil War had just ended.

She and her mother landed at Castle Garden and walked up Broadway to City Hall, with bundles of clothes and pots and featherbeds in their arms. The singing of the then exposed telegraph wires frightened them, as did the bustle of the people in the streets. They lost their fear when they met an Irish policeman who directed them to a rooming house on Baxter Street.

It must have been spring or summer when they arrived. The windows were open; and she was wakened often in the night by the sound of drunken voices singing:

We'll hang Jefferson Davis
On a sour apple tree.

For later Irish immigrants, the first sight of America was the Statue of Liberty. Elizabeth Phillips, who arrived in 1920, remembered it well.

The first time I saw the Statue of Liberty all the people were rushing to the side of the boat. "Look at her, look at her," and in all kinds of tongues. "There she is, there she is," like somebody who was greeting them.

Eighteen-year-old Francis Hackett described what he saw when he leaned over the rail of the ship Furnessia *in 1901.*

The first man I saw on the New York pier was a black-skinned human being. He was out in front of the shed, waiting to catch a hawser from the *Furnessia*. The friendly sunlight dwelt on him—it was October 6, 1901—as he stood by the bollard in his faded light blue overalls, gazing up for the rope that was to be cast to him. America was his fate. It was going to be mine, so we would be having the country in common.

That was a big surprise to me—a colored man. I knew of them, since one had come with a circus to Kilkenny. But here was this one, easy and free, giving a hand to us to land in his country. That was something new. It was all going to be new and different....

My heart had risen at the sight of land that morning. October had brushed the dying leaves with old gold, and the dry earth was astonishingly seared after a long summer. It must have been on Staten Island I saw the frame houses. I was used to stone houses and brick houses, or to thatched cabins with mud walls. It was marvelous to see such differences. The sky over New York was flawless, and it seemed much further away than the vaporous Irish sky. The warmth of the air was too much for a youth wearing a thick Irish tweed suit, but it was part of the novelty. Everything was like a vacation; it was all so gay and foreign. Yet under the excitement, the contrast that refreshes the imagination, I was aware, while grinning my way into a new family, that Ireland was forsaken. It was ten years to the day since Parnell [the Irish nationalist] died. Now I was entering into fraternity with another people, the white man and the black man.

Michael Kinney crossed the Atlantic in 1930 and recalled:

I come by boat, and I come [by train] right here to Pittsburgh, because that's where my brother and sister were, and I had aunts and uncles here. Stayed with my aunt. And if I had the money at the time, after a couple of weeks, I'd go back home again. I was so lonesome. I was all right until the evening come, and oh, my God, I used to long for that home. I used to cry, I may as well tell you. We used to dance at the crossroads at home and go out every night and have a good time. And it's a whole year to get to know anybody here. Yeah, I was lonely for a whole year, but then I settled down.

Everything surprised me. Everything was different. Different food—well, it's not the food alone. Back there we never used to ride, we always walked. We never went to ride a bus— we were lucky we had a bicycle. And I had to get used to electric lights. I was always trying to blow that light out when I came here. And no gas and no water on the farm. You were outside and brought water in. The fire was there and everything was cooking. We used to cook a big pot of potatoes, you know. We used to eat five or six potatoes at home. And I seen my aunt here cooking a couple of potatoes, and I thought to myself, "Why is she cooking them couple of potatoes? That's all they eat here, one or two potatoes?" But I got more to eat here than I got back there, you know.

A 19th-century cartoon shows the traps that new immigrants might fall into. Swindlers of all kinds were ready to separate the newcomers from their money.

In 1853 the Illustrated News of New York *told the story of an imaginary immigrant, Paddy O'Dougherty. He arrived penniless in New York, but soon became a stevedore. Working on bridges and the railroads at 40 cents an hour, he saved his money and brought over his wife and children:*

And now pursuing the even tenor of his way, minding his own business, working hard, but his wages being gradually raised, we see Paddy fairly afloat in the New World. Six months after, Paddy remitted home the money for the passage of his wife. Nine months from the day he landed he moved his wife and family into an humble little dwelling, entirely paid for by his earnings.... Eight years from the time he first engaged himself to Timothy Brown at forty cents a day, Paddy O'Dougherty was worth one hundred thousand dollars, owned a country seat within sight of his first day's labours, and employed Timothy Brown as superintendent of it at two dollars a day; and, though rough be the name, few gentlemen more affable, polite, or intelligent, or more gentlemanly in appearance, can this day be seen in Water Street than this same Paddy O'Dougherty.

A New York City census taker in the Fourth Ward—a neighborhood of Irish immigrants—in 1879. The newspaper cartoonist depicted the hostile reception that immigrants gave to city officials.

SETTLING IN

Jim Doyle, who immigrated to the United States in 1817, before the great migration, landed in Philadelphia. But after three months of working as a printer, he left for New York, where he found a job selling maps for a bookseller. Within a year, he was in business for himself selling a stock of pictures. On January 18, 1818, he wrote home to his wife.

I am doing astonishingly well, thanks be to God, and was able on the 16th of this month to make a deposit of 100 dollars in the bank of the United States....

[Here] a man is allowed to thrive and flourish without having a penny taken out of his pocket by government; no visits from tax gatherers, constables, or soldiers; everyone at liberty to act and speak as he likes, provided he does not hurt another; to slander and damn government, [and] abuse public men in their office to their faces.... Hundreds go unpunished for crimes for which they would be surely hung in Ireland; in fact, they are so tender of life in this country, that a person should have [great difficulty] to get himself hanged for anything.

Many immigrants left families behind, intending to bring them over to America with their first earnings. Thomas Garry from County Sligo arrived in Canada at St. John's, New Brunswick, in 1847 and immediately went south to the United States. He wrote to tell his wife about his railroad job and promised to send money.

Be on the watch at the Post office day after day I wont delay in Relieving yous as it is a duty Encumbered on me by the laws of Church and I hope God will Relieve me...I long to see that long wished for hour that I will Embrace yous in my arms there is nothing in this world gives me trouble but yow and my dear Children whom I loved as my life....

You will shortly be in the lands of promise and live happy with me and our children.

No more at Preasant

From your Faithful husband till death
Thos. Garry

[P.S.] I was ready to go to work to pay Passge for you and the children but i consider yous would not stand the wracking of the sea till yous be nourished for a time.

A young worker in Philadelphia wrote to his uncle, a farmer in Ulster, in 1854.

I have got along very well since I came here and has saved some money. I never regretted coming out here, and any young person that could not get along well there would do well to come here, if they intended to conduct themselves decently.... But old people have no great chance here.

In 1979 Mary Ann Kelly retold the family legend of how her great-grandparents came to settle in Kentucky.

In 1854 the boat from Ireland left Galway's Ballygaddy Road far behind, bearing our great-grandparents Martin and Margaret Hart Kelly, with our grandfather as a little boy and his sister, Bridget. It was blown off course before landing on the east coast and so arrived farther south than intended. Then the Kelly family had to come up the river by steamboat. The captain wanted our great-grandmother to name her unborn child after his boat. But she didn't. She called him Andy instead of "Belle!" Attracted to the land of Kentucky, they settled in Ashland in the Iron Mountain region on the Ohio River.

In 1873 a Catholic priest, Father Stephen Byrne, published a book of advice for Irish who were considering immigrating to the United States. Byrne warned against settling in the large cities.

The high rents and high prices of the common necessaries of life are [drawbacks]. Thus, for instance, in the city of New York, no laborer or mechanic can get a decent room or two in a *tenement-house* under twelve or fifteen dollars a month. Such persons...can hardly be expected ever to rise to independence. It is clear and undeniable that men of the same class have gone either to the smaller towns of the West and South, or to the country parts, and have acquired their own homes in every case in which steadiness in work and sobriety justified the hope of their doing so.

Katherine O'Hara described her first days in the United States.

I arrived at Boston on May 8, 1930. My aunt was not at the pier to meet me. The immigration authorities took over. They were very kind. They asked me how much money I had to pay for a taxi, and I told them—I had earned the money on board the ship coming over. They took me to Commonwealth Avenue, where my aunt worked as a cook for a millionaire family. When we got there, the caretaker said that my aunt had moved on to the summer residence, but she had arranged for me to stay in a rooming house.

A New York City Irish neighborhood on Washington Street, near the Hudson River docks, around 1890.

Paradise Square, better known as the Five Points, which was the heart of the Irish immigrant section of New York City in the 1850s. The poverty and crime here made it a notorious slum.

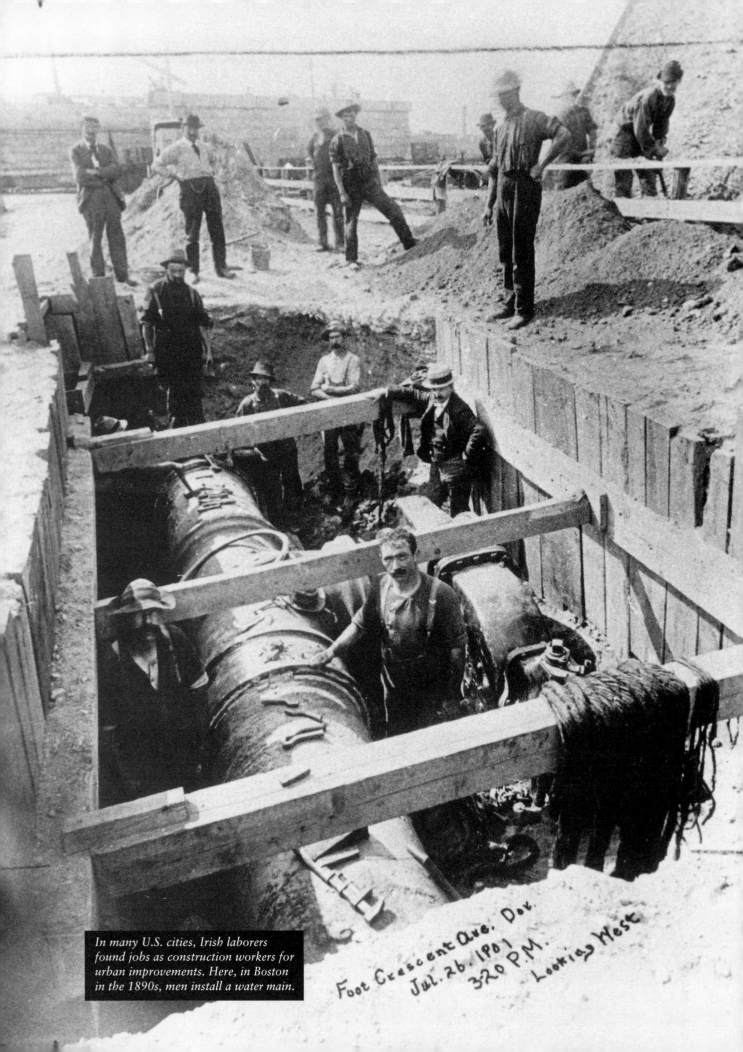

In many U.S. cities, Irish laborers found jobs as construction workers for urban improvements. Here, in Boston in the 1890s, men install a water main.

Foot Crescent Ave. Dor.
Jul. 26. 1901
3.20 P.M.
Looking West

CHAPTER FOUR

A NEW LIFE

The Irish immigrants between 1781 and 1820 tended to be artisans, tradespeople, teachers, merchants, and professionals. They found a ready market for their skills in their adopted country, and many prospered. Irish Protestants established cloth-making mills in the four New York counties that today bear the names Ulster, Orange, Greene, and Sullivan (the last named for an Irish American general of the American revolutionary army). Thomas O'Connor, a former member of the United Irishmen, founded *The Shamrock,* the first Irish American newspaper, in 1810. John Doyle, another immigrant, wrote home in 1818, recommending the United States as a "free country, where a man is allowed to thrive and flourish without having a penny taken by government."

However, the growing numbers of poor Irish Catholics who began to arrive after 1820 aroused fears among nativist Americans. The predominantly Protestant population of the United States had a deep-seated prejudice against Catholics, often disparagingly called "papists" because of their allegiance to the pope in Rome. Of course, Catholics had come to

America from early colonial times, but most had been English or French who practiced their religion quietly, and some adapted to their new country by becoming Protestants. The Irish Catholics who arrived after 1815 asserted their right to establish churches and parochial schools, becoming a visible presence.

The prejudice against Irish Catholics sometimes took violent forms. In 1834 a mob in Charlestown, Massachusetts, burned down a Catholic convent where nuns operated a school. Ten years later, riots broke out in Kensington, an Irish neighborhood in Philadelphia, when Protestants gathered to protest the use of the Catholic version of the Bible. Deadly fighting between Catholics and Protestants continued in Kensington for nearly two months.

Irish Catholics, hardened by centuries of British oppression, were not shy about defending their rights. When rumors spread through New York City that anti-Catholic groups planned to repeat the violence that had devastated Kensington, the Catholic archbishop of New York, Irish-born John Hughes, took action. Hughes sent Irish volunteers to guard every Catholic church in the city. He warned that he would order his volunteers to burn New

York if even one church was damaged. None was.

After the famine in Ireland pushed even greater numbers of Catholics to the United States, the anti-Catholic and antiforeign prejudice increased. In 1845 a new political party was founded by "native" Americans who hoped to block foreign immigration. Officially named the American party, the group was dubbed the Know-Nothings because its members responded to questions about their activities by saying, "I know nothing." In 1854 the Know-Nothings reached the height of their influence by electing governors or a majority of the legislatures in seven states.

It was also around the 1840s that Protestant Irish immigrants began to call themselves Scots-Irish to distinguish themselves from the predominantly Catholic newcomers. The refugees from the famine were different in many ways from the Irish who had made up the first wave of immigrants. Mostly illiterate, they had no skills except farming. It would have seemed natural for them to take up the same occupation in the United States, but in fact relatively few did, primarily because they lacked the financial resources to become homesteaders.

From the 1820s on, most Irish

immigrants found their first jobs in the United States as laborers. They were among the workers who built the Erie Canal, the transcontinental railroad, and the Statue of Liberty in New York Harbor. Irish laborers enriched the great cities in which they lived by helping to construct their roads, bridges, elevated railways, subway systems, and skyscrapers. John Daniel Crimmins, the son of an Irish immigrant, started as a hod carrier and worked his way to the ownership of his own construction company in New York City. Crimmins's firm completed more than 400 buildings, miles of streets and gas lines, and much of New York City's elevated railway system.

The California gold rush, which began at the height of Irish emigration in 1849, attracted numerous hopeful prospectors from the Emerald Isle. As more discoveries of precious metals were made in the western states, Irish laborers filled the need for miners.

A fortunate few became millionaires from the mining industry. James C. Flood, a San Francisco saloon keeper, invested in a Nevada silver mine with three other Irish immigrants. Flood and his partners controlled the Comstock Lode, the largest deposit of silver ever found in the United States. With their shares of the wealth, the four men and their families entered society in San Francisco and New York, becoming known as the "Irishtocracy."

"Leadville Johnny" Brown had

a similar success in the gold and silver mines of Colorado. Brown and his wife, Margaret Tobin, built a lavish mansion in Denver. Margaret became a legendary figure known as the "unsinkable Molly Brown" when she survived the sinking of the ocean liner *Titanic* in 1912.

On a Boston wharf in 1882, the wife of one of these Irish clam diggers has brought lunch for her husband.

Irish immigrant women often found work as servants in the houses of the upper class. Castle Garden had a labor exchange where well-to-do families chose servants from groups of newly arrived Irish women. James Flood's wife and Johnny Brown's mother had both been housemaids. The Irish "servant girl" was so familiar that she became a stock character in 19th-century plays and inspired cruel ethnic jokes.

Single women—and children, too—also worked in the cloth mills of New England and took other kinds of factory work that required

long hours of repetitive labor.

Both male and female Irish immigrants continued to encounter prejudice that kept the majority at the bottom of society. Newspaper advertisements sometimes carried the line "No Irish need apply."

Yet the Irish, concentrated in large cities, took advantage of the fact that the men could vote. Their numbers made them a powerful political force. Having learned the techniques of organization from the successful campaigns of Daniel O'Connell in Ireland, the Irish in America proved to have a genius for politics. In the cities where they were most numerous, such as Boston, New York, and Chicago, the Irish took control of local, tightly controlled political organizations, called machines, in the latter part of the 19th century.

The best-known example is the Tammany Hall machine in New York City, named for the headquarters of the local Democratic party. "Saint Tammany" eventually commanded an Irish loyalty second only to Saint Patrick. When William "Boss" Tweed (who was not Irish) became leader of Tammany Hall in 1868, he distributed the spoils of victory among his Irish supporters. The Tammany machine continued its dominance over New York politics under a succession of Irish leaders until the 1940s.

In other cities, Irish American politicians also gained power. Hugh O'Brien became the first

Irish mayor of Boston in 1886, beginning a tradition that included such colorful figures as James Michael Curley and John F. "Honey Fitz" Fitzgerald, grandfather of President John F. Kennedy.

In Chicago the Irish machine started later but lasted longest. Two sons of immigrants, Michael "Hinky Dink" Kenna and "Bathhouse John" Coughlin, dominated Chicago's heavily Irish First Ward from around 1895 to the 1930s. Their support was crucial to the election of Democrat Anton Cermak as mayor in 1931. After Cermak's death, the Irish-dominated Democratic party of Cook County elected Irish American mayors for the next half century. The Chicago machine reached the height of its influence under Mayor Richard J. Daley, who served from 1955 until his death in 1976.

These political machines worked because they met the needs of the voters in a very direct way. Through a network of neighborhood leaders, the machines offered money, jobs, and even food when someone in their territory needed it. The grateful recipients returned the favors with their votes.

Well into the 20th century, newcomers from Ireland—and later generations of Irish Americans—obtained jobs through the urban political machines. A note from the local leader was usually enough to get a man a job in the police or fire department or with a construction company doing business with the city. Though the Irish cop, like the Irish maid, became the butt of jokes and the object of satire, his presence in every neighborhood was a sign of the Irish

Irish immigrant James J. Carney founded this basket factory near Lawrence, Massachusetts, in 1875. The style of weaving he used originated in Ireland.

immigrants' political power.

Irish Americans became a force to be reckoned with in national politics as well. In the closely contested Presidential election of 1884, both parties hoped to win the electoral votes of the nation's largest state, New York. A few days before the election, a Protestant minister who supported Republican candidate James G. Blaine branded the Democrats as the party of "rum, Romanism, and rebellion." This statement—a reference to the supposed drinking habits of the Irish, their religion, and their resistance to British rule of their homeland—propelled irate Irish voters to the polls. Grover Cleveland carried New York by only 1,200 votes, becoming the first Democratic President since 1861.

The Irish talent for organization extended to their working life as well. Irish Americans were active in the struggle to organize laborers into unions. In 1879 Terence Powderly, the son of Irish immigrants, became head of the Knights of Labor, a national association of labor unions. Under his leadership, the Knights expanded to more than 700,000 members.

Among the heroes of the labor movement were such Irish American women as Mary Harris "Mother" Jones, born in County Cork. Women's involvement was only natural, for they worked as hard as the men, and their Irish sense of justice was as strong.

In the 20th century, many Irish Americans have served as union presidents. George Meany, who began work as a plumber's apprentice at age 16, became the first head of the combined American Federation of Labor–Congress of Industrial Workers (AFL–CIO), the nation's largest labor organization, in 1955.

Six years later, John F. Kennedy became the first Irish Catholic President of the United States. His election, like George Meany's, represented the triumph of generations of Irish Americans who had arrived here with nothing but faith in the American dream.

I'm a decent lad just landed from the
     town of Ballyfad;
I want a situation and I want it very bad.
I've seen employment advertised. "It's just
     the thing," says I,
But the dirty spalpeen [rascal] ended with
     "No Irish Need Apply."
"Whoo," says I, "that is an insult, but to
     get the place I'll try."
So I went there to see the blackguard with
     his "No Irish Need Apply."

Chorus:
Some do think it is a misfortune to be
     christened Pat or Dan,
But to me it is an honor to be born an
     Irishman.

I started out to find the house, I got there
     mighty soon;
I found the old chap seated—he was
     reading the *Tribune.*
I told him what I came for, when he in a
     rage did fly;
"No!" he says. "You are a Paddy, and no
     Irish need apply."
Then I gets my dander rising, and I'd like
     to black his eye
For to tell an Irish gentleman "No Irish
     Need Apply."

I couldn't stand it longer so a-hold of him
     I took,
And I gave him such a welting as he'd get
     at Donnybrook.
He hollered "Milia Murther," and to get
     away did try,
And swore he'd never write again "No
     Irish Need Apply."
Well, he made a big apology; I told him
     then goodbye,
Saying, "When next you want a beating,
     write 'No Irish Need Apply.'"

PREJUDICE

In the early 19th century, anti-Irish and anti-Catholic prejudice ran so high that priests were often attacked. One priest in Portland, Maine, described his fear in 1854.

Since the 4th of July I have not considered myself safe to walk the streets after sunset. Twice within the past month I have been stoned by young men. If I chance to be abroad when the public schools are dismissed, I am hissed and insulted with vile language; and those repeated from children have been encouraged by the smiles and silence of passersby. The windows of the church have frequently been broken—the panels of the church door stove in, and last week a large rock entered my chamber unceremoniously about 11 o'clock at night.

Anti-Irish and anti-Catholic violence was whipped into a frenzy in Philadelphia in 1844 as a result of Catholic schoolchildren being allowed to read the Catholic version of the Bible. Mobs burned churches and ran amok through Catholic neighborhoods. A Catholic nun who worked in an orphanage wrote about her fear to her superior in a letter of May 9, 1844.

We are in the midst of frightful dangers; a great portion of our peaceful city is the scene of a dreadful riot and bloodshed. Two of our Churches [are] burned to the ground...St. John's has been guarded since Monday night, and St. Mary's is now surrounded by a strong detachment of the Military.... The clergymen have left their dwellings, the Bishop his house, the priests and students have deserted the seminary, everyone seeking a night's lodging in the family of some friends. Three police officers now guard our asylum, and we know not what moment our dear little ones must be roused from their peaceful slumbers, to fly for their lives.

The Know-Nothings of the antiforeign American party had their greatest strength in Massachusetts. The party's candidates swept the state in 1854. The Know-Nothing–controlled legislature set up the Smelling Commission to inspect convents in a search for scandal. In March 1855 the commission investigated the Notre Dame Academy, a boarding school. A nun recalled:

Soon the expected committee came, twelve in number, headed by no less a personage than the mayor of the city. According to their pastor's instructions, the Sisters refused them admittance until they saw Father Timothy who escorted them through the house, asking them whether they met the extraordinary sights they had expected. They insisted upon all the closets being opened for inspection, which was accordingly done, the children's dormitories were visited, and lest anything should escape observation the worthies...examined the beds. When, however, they were about to enter the dormitories of the religious, the Reverend Father forbade them to cross the threshold as they valued their own safety. The twelve desisted and in taking their leave expressed themselves satisfied with the result of the visit. Neither mayor nor committeemen made his appearance at Notre Dame again.

The Irish often fought back against the Know-Nothings. In 1854 when mobs tried to attack St. Patrick's in Lowell, Massachusetts, the Irish fiercely defended their church. The June 1854 entry from the annals of St. Patrick's Convent described the incident.

Just at dusk one quiet evening, the ominous peal sounded forth from the belfry. Fear and consternation were in the many hearts, but trustful prayer in the little convent. The self-constituted defenders stood with arms uplifted, ready to hurl their missiles at the first assailant. Yes, the Know-Nothings were approaching the church, but they had not counted sufficiently on Irish loyalty and vim. When just within sight of St. Patrick's, they were attacked by some strong-armed Irish men and women,—yes, women for these led the attack. The march became a melee, and the street was completely filled with the motley crowd. They reached the bridge, which spans the canal just within sight of the convent. There was a halt, a splash and a ringing cheer—A sinewy matron unable to re-

The anti-Catholic and anti-immigrant Know-Nothing party flourished for a time in the 1850s. The North Carolina Weekly Standard *published a joke menu from a Know-Nothing banquet:*

Catholic broth	Jesuit soup
Roasted Catholic	Broiled priest

The Pope's Big Toe, broiled

Fried nuns, very nice and tender

(Dessert) Rich Irish Brogue   Sweet German Accent

An artist depicted the anti-Catholic violence in Kensington, Pennsylvania, in 1844. The men in tall beaver hats are self-styled "native Americans" firing at state militia defending a Catholic church.

55

The circus exhibit in this 1891 cartoon showcases "The only policeman ever born in America." It pokes fun at the fact that the police in cities such as Boston, New York, and Chicago were frequently Irish immigrants.

strain her indignation had seized upon one of the leaders of the gang, and flung him over the railing foundering into the water below. The rest of the gang made the best of their way out of the mob, and although the Sisters were still in a state of anxiety, yet the attitudes of their assailants grew less and less threatening.

William O'Connell, born in Lowell, Massachusetts, in 1859, would one day become a cardinal of the Roman Catholic church. But his first school days were spent in a New England public school where the teachers were prejudiced against Irish Catholics.

My teachers of that time were of the strictest Puritanical mold, with the coldest idea of duty as they saw it...with a suspicion and distrust of everything Catholic.... We [the Catholic students] lived actually in an atmosphere of fear. We sensed the bitter antipathy, scarcely concealed, which nearly all of those good women in charge of the schools felt toward those of us who had Catholic faith and Irish names. For any slight pretext we were severely punished....

At the time I was serving at the altar and I remember absenting myself so as to assist at the Mass on Good Friday. There were two or three other Catholic boys in my class who did the same. Immediately on the opening of the class on Saturday morning, the teacher, a rabid anti-Catholic,...angrily demanded an explanation of our absence the day before. I answered that I was absent to be present at Mass, for yesterday was Good Friday. With a sneer, she told the whole class that that was no reason whatsoever. "Good Friday!" she said, "I want you to know that all Fridays are good, and the boys who absented themselves may now come forward for punishment." It can easily be imagined that a school filled with an atmosphere like that was certainly in no way conducive to the pleasures of learning.

James Michael Curley, who would one day be Boston's mayor, was a bitter foe of the Boston Brahmins, as the upper-class Protestants were called. Curley wrote in his autobiography:

The Brahmins] spoke contemptuously of the "cattle Irish," who, packed together like animals, came in steerage on ships...from Queenstown to Boston around the mid-nineteenth century. The colleens who found jobs in the kitchens of the wealthy were called "pot-wallopers," "biddies," and "kitchen canaries," and sometimes they rebelled. There is the story of the Back Bay maid who served a Thanksgiving turkey with one leg missing. Fired when she explained that she gave it to the cop on the beat, she picked the turkey up by the other leg and threw it at the dowager who had called her a "dirty Irish pig."

"I'm not fired," she said. "I quit."...

In my mother's day, even a job as a domestic servant was not always open to Catholics, for the simple reason that well-to-do Protestant employers did not wish to be bothered with taking them to church on Sundays in horse-drawn carriages.

Irishmen, who often brought picks and shovels with them, were forced into jobs as day laborers, hostlers, stablers or waiters.... Other "greenhorns," "clodhoppers," "Micks" or "Paddies," as the Brahmins called them, worked as ditchdiggers on the canals and sewers, or as laborers on railroads while they lived in barracks. Still others found employment in a laboring capacity with the city. Those who shoveled muck were labeled "muckers," and this term was applied, as by Henry Adams, to the sons of Irish immigrants.

In their book I was a "mucker."

"Micky M'Carty is Rising in the World, slowly, but surely—

Sometimes prejudice stood in the way of getting a job. Rose Lynch, born in Dublin in 1906, one of 14 children, immigrated to the U.S. when she was 21. She first settled in New York City.

In New York you had loads of employment agencies. Often the job listings stated, "No Irish Need Apply." So we went only to the ones that didn't say this. I got a job making hot dogs in a meat factory.

Even those who found work experienced hardships because they were foreigners. An Irish immigrant who found a job in the southern states working on a railroad wrote home about the conditions.

Sister Norah is staying at one of the Big Hotels—

It would take more than a mere letter to tell you the despicable, humiliating, slavish life of an Irish laborer on a railroad in the States; I believe I can come very near it by saying that everything, good and bad, black and white, is against him; no life for him—no protection in life; can be shot down, run through, kicked, cuffed, spat on; and no redress, but a response of, "Served the damn son an Irish b—— right, damn him."

Ould Mr. Fogarty has been called to the Bar—

In 1873, a U.S. newspaper published this series of cartoons, called Letters to the Ould Country. *The titles are supposed to be from an Irish immigrant's letter home, describing in grand terms what is in reality poverty and hard work. "Called to the Bar," for example, usually referred to becoming a lawyer, but the picture shows a man who has been arrested and hauled before a judge.*

As for us, we are living on Fifth Avenue, near the Cinthral Park—

The immigrants who came to America expecting a better life were sometimes disappointed. A young immigrant who arrived in Washington, D.C., in 1851 wrote his brother the following year:

For myself I am now in a fair way bettering myself, but I will tell you...that I have suffered more than I thought I could endure, in a strange Country far from a friend, necessitated to go on public works from four o'clock of a summer morning until eight at night enduring the hardships of a burning sun, then sickness losing what I dearly earned for my short time in this country. I have experienced a great deal, which may serve me the remainder of my life.

Irish immigrants were willing to work hard to get ahead. Many labored in mines and factories. This photo shows the blast furnace of the Lackawanna Iron and Coal Company in Scranton, Pennsylvania, around 1885.

GOING TO WORK

The 19th-century Irish immigrants often worked in the most dangerous jobs. In the 1830s many of them helped build a canal that connected Lake Pontchartrain with New Orleans. Irish-born actor Tyrone Power (great-grandfather of the 20th-century movie star) described the scene.

The hundreds of fine fellows [labored] beneath a sun that at this winter season was insufferably fierce, and amidst a pestilential swamp whose exhalations were fetid to a degree scarcely endurable for a few moments; wading amidst stumps of trees, mid-deep in black mud, clearing the spaces pumped out by powerful steam engines; wheeling, digging, hewing, or bearing burdens it made one's shoulders ache to look upon; exposed meantime to every change of temperature in log huts laid down in the very swamps.... Here they subsist on the coarsest fare...often at the mercy of a hard contractor, who wrings his profits from their blood; and all this for a pittance that merely enables him to exist.

William O'Connell, later to become a cardinal of the Catholic church, was born in 1859 into a large Irish family. His parents lived in a middle-class neighborhood, but William, as a teenager, took a summer job at the mills.

For the first hour or two I had the greatest delight in scattering the fluffy cotton mass on to the table of the carding machine and watching it with fascinated eyes pass through the machinery, come out in great bands of soft white, down stuff, automatically twist itself around the huge

bobbin...[but] the thought came to me, "What if I were condemned to do this all my life!" And the poetry of mechanical motion began to turn into very serious prose. I began to feel faint from the disgusting smell of the oil, whose vapor filled the atmosphere of the room. I looked around me and saw my boyish companions, who at that moment looked to me like shriveled old men.

The stench of ammonia that rose from the chemical room near-by made my eyes run water and almost stifled my breath. I suddenly felt a weakness as if I were about to faint. I stopped the machine and leaned against it in an endeavor to recover. Somehow I felt I could never start that machine again. [William O'Connell quit during the lunch break.]

David Lawlor, who wrote his autobiography in 1936 when he was 71, described the horror of a fire in one of the cloth mills in Fall River, Massachusetts.

When I was nine years old I witnessed a great tragedy. I saw the burning of the Granite Mill No. 1, where twenty-eight men and women were burned to death. I was in No. 2 across the street on the nineteenth of September, 1874. I saw a puff of smoke coming out of the window and a man descending the fire-escape. The fire was in the mule room [where the machines called "mules" spun fiber into thread].... The only way out was a big door which led to the tower. This door was said to have been locked.... The neighbors rushed mattresses at each end of the mill and some of the boys and girls rushed to the window, said a prayer and jumped. One of those little boys had been a companion of mine in the mill and was sent over to No. 1 a few days before. He jumped, ran home, told his mother the mill was on fire, and dropped dead. Few escaped from that room.

One woman, a Mrs. Murphy, lost three daughters. It had been her custom to bring their breakfast every morning, but that morning of the fire she was in Providence. After the fire for some time she would get the breakfast, fill the pails, walk down to the ruined mill and cry hysterically until some friends would take her home again.

Irish-born labor leader Mother Jones described the conditions she saw as a young woman in a southern factory.

This factory was run also by child labor. Here, too, were the children running up and down between the spindles. The lint was heavy in the room. The machinery needed constant cleaning. The tiny, slender bodies of the little children crawled in and about under dangerous machinery, oiling and leaning. Often their hands were crushed. A finger was snapped off.

A father of two little girls worked a loom next to the one assigned to me.

"How old are the little girls?" I asked him.

"One is six years and ten days," he said, pointing to a little

Young Irish women—many of whom came to America alone—had to find work to support themselves. Child's restaurant, on 42nd Street in New York, hired "respectable" young women as waitresses.

My parents live in Ireland and are entirely dependent on myself and sisters for support. I served an apprenticeship of three months, seven years since. I have worked at the trade ever since. I am a good seamstress and work hard. I try but I can not make over $1 per day. I pay rent for machine, $2.50 per month. Am not able to afford to ride on street cars, therefore I have to walk, and if I happen to be one minute late, I have to walk up long flights of stairs and am not allowed to go on the elevator.

Factory work was one way Irish immigrant women could earn a living. The Kimball Shoe Company in South Lawrence, Massachusetts (shown here in 1903), employed both women and men in conditions that were relatively good for the time.

girl, stoop shouldered and thin chested who was threading warp, "and that one," he pointed to a pair of thin legs like twigs, sticking out from under a rack of spindles, "that one is seven and three months."

"How long do they work?"

"From six in the evening till six comes morning."

"How much do they get?"

"Ten cents a night."

"And you?"

"I get forty."...

They die of pneumonia, these little ones, of bronchitis and consumption. But the birth rate, like the dividends, are large and another little hand is ready to tie the snapped threads when a child worker dies.

For most immigrants and second-generation Irish, life was an endless round of hard work. Jobs were often temporary, and workers moved from place to place. One Irish American told his story to an interviewer in the 1930s.

I was born May 17, 1888, the year of the big Blizzard. In my home town—Nashville, Tennessee—the snow was up to people's necks. My father died when I was three months old, and I was raised in an orphan asylum until I was ten. Then I went to work in a bucket factory to support my mother. I was twelve years old then.... At sixteen years of age, I left home and started working in boiler shops wherever I could find work. I rode freight trains all over the country. I went through Oklahoma and Texas and worked on bridges and railroads as a structural ironworker. Then, in Port Arthur, Texas, I went to work for the Texas refinery. I served out my time as a boilermaker and got my card. Then, I went into the oil fields during the oil boom of 1912 and 1913, building oil storage tanks.

When the war [World War I] broke out, I went to Virginia and worked for the Chicago Bridge and Iron works...and helped build a steel water tower, one of the largest in the world. It was built to prevent fire....

After completion of the job, I obtained a job as boilermaker in the Hog Island shipyards, the largest shipyards in the world. Fifty ships could be built there at one time. I worked there until the armistice. When the armistice was signed, it closed overnight.

From there I came to Newark and obtained work in the federal shipyards as a boilermaker. After working for two years, I went into the steeplejack business for myself, and I have been working in that business since—eighteen to twenty years. I had two falls. One fall was a hundred feet. I fell off a smokestack at Napthea hat works, William and Main streets. After being laid up for three months, I had a job at Haledon, New Jersey, for the Manley Piece dye works and fell fifty feet off of a stack. Yet I have no defect from either fall.

Now I find it hard to get a job, and I have to take work as a watchman.

Michael Donohue, the son of immigrant parents who settled in New York around 1905, described how he decided to take a civil service test for city employment.

I wanted to be an artist, but I didn't feel that it was in the cards right then. It was too insecure for someone from my background. I was concerned about finding a gainful occupation.... The only great desire I had was to become a civil servant and I zeroed in on that. It was the only idea I got encouragement for. And essentially, for the typical Irishman, that was the crux of what to be in those days. For one thing, it required very little in the way of education....

[Donohue passed the civil service test for a fireman's job.] I wanted to be the best fireman in the city—that's the kind of conceit I had—and so I wanted to go to the busiest firehouse, which at that time was Hook & Ladder 26, on 114th Street and Madison Avenue. We used to respond to about 300 fires a month....

It got to be that you could respond to a fire alarm automatically. Before you went to sleep, you'd set your boots and pants in a certain way, so that you could move right into them. When the bells started ringing...you'd be moving even before you were awake. You'd slide down one pole waking up on the way down, and then run over to the next pole and slide down it. In those days, the firehouses were built two or three stories high. If you were a little careless, you could break your legs. One guy in my firehouse was killed coming down....

If you made it down the poles okay, then you had to worry about collapses in the buildings where the fires were occurring. You could be going along and think God is on your side and suddenly you're laying flat. I've had ceilings collapse on me and suffered burns....

You'd go into an area and boy! The smoke'd be down to your shoes and you were lucky to get out of those situations. You'd be on the stairway and you'd have to go above the fire, where somebody could casually open a door below you by mistake and the whole thing's just a furnace.

Work was hard even when the employer was the city. A Jesuit priest named Philip Carey, born into an Irish family in 1906, described his father's job in New York City.

My father worked as a conductor on the 23rd Street trolley line. See, if you were literate, they made you conductor. Otherwise, they put you on the front end of the car. He was off two days a year. We used to see him Sunday afternoon, when he let us ride on the front of the trolley....

There was a trolley strike in 1916. During the strike, he came down to the waterfront to get some work. He worked the barbed wire and the cement, because no one would handle that. The breathing was awful. The barbed wire just tore all his clothes. There was a wonderful priest at that time at Guardian Angel, Father McGrath, who told my father after two days,

Many young Irish women worked as seamstresses. One of them glowingly wrote home:

I am getting along splendid and likes my work...it seems like a new life. I will soon have a trade and be more independent...you know it was always what I wanted so I have reached my highest ambition.

In cities where Irish Americans had political connections, immigrants obtained jobs through friends in government. Jeremiah F. O'Leary was the driver of Engine Wagon No. 7 of the Lawrence, Massachusetts, Fire Department around 1900.

"You're not the man for this kind of work. Get back. Don't stay around here."

Soon, we were down to having tea and toast. Every week the superintendent would come around and beg my father to come to work. Then they brought in all these scabs from the Bowery. They didn't know anything about the operation of trolley cars. But we knew it was over when they brought in all the fellows from out of state.... When the strike was over, my father went back to work, but he always wore his union button inside his hat....

He was a man with an enormous sense of justice. God would love him. A normal man would have been bitter at the treatment he was getting. He wasn't that way. He would say, "I am thankful to God that I will not have to face my Judgment Day with that on my soul. They will have to do it. I won't." He was always that way. When he was makin' enough money to pay his first income tax, he said, "Heretofore, I've been a freeloader. Now, I'm beginning to pay my own way."

Paul O'Dwyer, who arrived in the United States in 1925, had the support of his four older brothers—Bill, Frank, Jack, and Jimmy—who had immigrated before him. Nonetheless, as O'Dwyer later wrote in his autobiography, he had trouble finding a job he could hold for long.

On April 21, 1925, I arrived in New York with twenty-five dollars in my pocket, a single suit of clothes—on my back—and a straight razor....

Two days later I went to work in a garage. Bill had spoken to a precinct captain and the captain had recommended me. [Already active in politics, Bill O'Dwyer would later be mayor of New York City.] My job was to distribute automobile parts to the mechanics, but the parts confused me.... I had been there only a few days when I overheard someone refer to me as a "donkey," a derogatory term for an Irishman....

After about a month the boss said to me, "Look, kid, you're a nice kid, but you're really not suited to this kind of work," and that undid my career as a stock clerk. I found that the *World* had the best employment ads. I answered one for a stable boy, and the boss, assuming that I, being Irish, knew all about horses, gave me the job.... I lasted one day...bowing to Mrs. Maguire's insistence that I couldn't live in her boarding-house if I worked in a stable....

So again I consulted the *World* and found an advertisement for an elevator operator at 807 Riverside Drive. I met Dick Ryan, the superintendent, who was a County Clare man. He asked me how old I was, how long I had been in the country and what county I came from....

I loved the work...as long as none of my friends saw me on the job in my uniform. I did not want word to get home to Ireland that I was working as a "lift" operator.... In Bohola [his hometown] it was expected that my brothers by now would have sufficient prestige in America to convince a bank or simi-

Cloth mills in the northeastern states often employed children. It was thought that their small, nimble fingers could guide the thread into the machines more easily. However, children had to give up school and the opportunity for a better life to work in the mills.

lar institution to avail themselves of my talents. In fact, I too shortly yielded to the same kind of expectation—to become a lawyer. I determined to quit my job and go back to school.

Irish who had come earlier often helped newcomers find jobs. Such was the case with Michael Kinney, who went to Pittsburgh in 1930 during the Great Depression.

The first job I got was warehouse work.... That's the first job I had, unloading cars and loading trucks. I went down there, and there was a lot of people down there looking for work at that time. That was '30. And the boss had come out and he said, "No, nothing doing." I went down there for a whole week. "Nothing doing. Nothing doing." And this morning when he come out and said the same thing, he did like that to me [beckons] and he called me over and he says, "You've got an old-country suit."

I says, "Yes, I do, and that's the only one I have."

"When did you come over?"

"Two weeks ago."

"Well," he said, "don't spend no more carfare coming down here. You'll be the first man to get a job here, if there's an opening. Give me your telephone number."

And two days later, I got the job. He was an Irishman himself.

Michael Kinney also described working in a steel mill.

Well, when I started at first—the first day I started in the mill—they put me down the hot flues, you know, underneath the furnace. Cleaning the flue dust out of the furnaces. They got trenches about this wide, and you crawl back in them. I had to get a pair of safety shoes and a pair of gloves. They put me down underneath the furnace, and I was ready to die down there. Thought I'd never come up out of there. It was like being in hell. Well, I went to the labor shanty. I wanted my money, I wanted to quit. He said, "I'll give you a nice job tomorrow." So he put me on the outside working, and I stayed there. I was unloading cars—manganese, silicon, aluminum, all that stuff....

Then I got up to third-helping in the furnace. Third-helping ain't bad. You help in the furnace where they make the steel. They called us *slaggers*. The heat is coming out of the furnace, throwing stuff into it. Hot! Anything you do on the open hearth is hot. Third-helper-whatever is supposed to...throw it in there. They put aluminum in, they put silicon in. Fifty pounds, most of them are. You have to stand from here to there and throw it in....

And I seen a lot of people getting hurt in the mill. A lot of people getting killed in the mill, too. I seen an explosion. A couple of pits blow up. And there was a bulldozer, and I seen a man blowed out of it and his foot taken off. A crane man, and the crane blowed up on him one day.

Police Captain Edwin B. Dyer, born in New York City in 1837, was one of the many Irish Americans who supported his family and served his community as a member of the police department.

An anonymous Irish American miner brings his lunch pail to work. He would spend the next 12 hours underground, digging the coal that provided light, heat, and power for American homes and industry.

Wealthy American families of the 19th and early 20th centuries employed numerous household servants. Irish immigrants were preferred for such jobs, and working "in service" was the most common occupation for Irish American women at that time.

A soap manufacturer used this picture of "Mrs. McCarty," an Irish American servant, to advertise its product. The copy read: "Do I look tired? Is my back broke and the clock at seven? Not a bit of it. My wash was out before 12 o'clock. Kirkman's Borax Soap did it, and the clothes are as sweet and pure as the driven snow."

"IN SERVICE"

Many Irish immigrant women took jobs "in service"—as maids, housekeepers, cooks, and scrubwomen. Wealthy and even middle-class families in the 19th century often went to the immigrant stations to hire Irish women just off the boat. Many of the Irish "service" women regarded this type of work as more dignified than laboring in a cloth mill or some other kind of factory. In addition, much of their pay often went back to the old country to bring relatives over.

In the mid-19th century young Ann McNabb joined her sister Tilly, who was already living in Philadelphia and working in the home of a Mrs. Bent. Ann recalled that this was the beginning of her own family's success story.

When I got here Mrs. Bent let Tilly keep me for two months to teach me—me bein' such a greenhorn. Of course I worked for her.... Then I got a place for general housework with Mrs. Carr. I got $2 [a week] till I learned to cook good, and then $3 and then $4. I was in that house as a cook and nurse for twenty-two years. Tilly lived with the Bents till she died, eighteen years.... How did we keep our places so long? Well, I think me and Tilly was clean in our work and we was decent, and, of course, we was honest. Nobody living can say that one of the McNabbs ever wronged him of a cent. Mrs. Carr's interests was my interests. I took better care of her things than she did herself, and I loved the children as if they was my own. She used to tell me my sin was I was stingy. I don't know. The McNabbs are no wasteful folk. I've worn one dress nine year and it looked decent then. Me and Tilly saved till we brought Joseph and Phil [their brothers] over, and they went into Mr. Bent's mills as weaver and spool boy and then they saved, and we all brought out my mother and father. We rented a little house in Kensington for them. There was a parlor in it and kitchen and two bedrooms and bathroom and marble door step, and a bell. That was in '66 [1866], and we paid $9 a month rent. You'd pay double that now. It took all our savings to furnish it, but Mrs. Bent and Mrs. Carr gave us lots of things. To think of mother having a parlor and marble steps and a bell! They came on the old steamer "Indiana" and got here at night, and we had supper for them and the house all lighted up. Well, you ought to have seen mother's old face! I'll never forget that night, if I live to be a hundred. After that mother took in boarders and Joseph and Phil was there. We all put every cent we earned into building associations. So Tilly owned a house when she died and I own this one now. Our ladies told us how to put the money so as to breed more, and we never spent a cent we could save. Joseph pushed on and got big wages and started a flour store, and Phil went to night school and got a place as a clerk. He married a teacher in the Kensington public school. She was a showy miss! Silk dress and feathers in her hat!

Joseph did well in his flour store. He has a big one on Market Street now and lives in a pretty house out in West Philadelphia. He's one of the wardens in his church out there and his girls give teas and go to reading clubs.

But Phil is the one to go ahead! His daughter Ann—she was named for me, but she calls herself Antoinette—is engaged to a young lawyer in New York. He gave her a diamond engagement ring the other day. And his son, young Phil, is in politics and a member of councils. He makes money hand over hand. He has an automobile and a fur coat, and you see his name at big dinners and him making speeches. No saving of pennies or building associations for Phil.

It was Phil that coaxed me to give up work at Mrs. Carr's and to open my house for boarders here in Kensington. His wife didn't like to hear it said I was working in somebody's kitchen. I've done so as to lay by a little sum every year. I heard that young Phil told some of his friends that he had a queer old aunt up in Kensington who played poor, but had a great store of money hoarded away. He shouldn't have told a story like that. But young folks will be young! I like the boy. He is certainly bringing the family into notice in the world. Last Sunday's paper had his picture and one of the young lady he is going to marry in New York. It called him the young millionaire McNabb. But I judge he's not that. He wanted to borrow the money I have laid by in the old bank at Walnut and Seventh the other day and said he'd double it in a week. No such work as that for me! But the boy certainly is a credit to the family!

Marie Haggerty's family went from Ireland to a New Brunswick farm. She left Canada to move to Boston in the second half of the 19th century. As a 72-year-old woman, she was interviewed in the 1930s about her experience as a housemaid.

It wasn't housework I did. I was a nursemaid or a second girl—never just an ordinary girl out to service. My aunts and uncle were very glad to have me working for such nice people, real high-class people. I had a good home and I was treated good. Now if I had gone into a factory to work, the folks would have been worried. The girls in the shops never made over six or seven dollars, and them that dressed so well on that, and paid their board, too, made people lift their eyebrows. I was lots better off. I got seven or eight dollars a week, my room, and it was always a nice one, and the best of food. I was really next thing to a lady's maid, for when the children went to bed, often the mistress would let me hook her dress, or brush her hair, and all the time she'd be talking to me, just like I was her equal.

I always had good jobs, and we usually worked by twos,

An Irish maid serves dinner to the Bellamy family of Boston. Even middle-class households could afford a cook or nurse-maid because the wages paid to servants were low. From the standpoint of the Irish immigrants, living "in service" with a respectable family was preferable to working in a mill or factory.

Irish immigrants who had learned lace making and sewing from their mothers found a ready market for their skills in the United States.

another girl and myself. A body didn't have to show references for jobs like they do now, but that wasn't the half of it. You got hired by your looks, and even if you looked honest, they would test you out. Once I was making up a bed, and right beside it was a five-dollar bill. I knowed nobody dropped that for nothing, so I didn't know if I should pick it up and tell them, or what, but my face burned like fire, for I knowed I was getting tested. I left it there all the time I worked in the room, and when I got done I put it on the bureau and put a vase over the end to make sure it didn't blow off. I was just going out of the room when the madam came in. I often think what would've happened if she'd come in while I was smoothing the bill out—would she believe I was going to put it on that bureau? I don't think so, for I was so new there. They often left food and fancy cakes around, just to test us, but I learned my lesson early on that. Once I just had my hand on a fancy cake in the parlor, and I got such a crack on the hand from the cook. She pulled me back to the kitchen and made me sit down and eat my fill of fancy cakes and told me never to take anything that was outside the kitchen, for it was always a trick to see how honest we was.

Irish women, like the men, were outspoken, even toward their employers. Elizabeth Dolan, who came to the United States at the age of 16 in 1912, described her service in a wealthy home in Boston.

I went to work for the Worthinghams, one of the ten oldest Yankee Brahmin families in Boston. I remember when Mrs. Worthingham died, there was two columns in the *Globe*. She was a millionairess.... She had a cook, which was me. She had a kitchen maid; she had a parlor maid; a butler, which was English right from England; a chambermaid; a lady's maid that used to travel with her on her trips; a laundress; and Stanley, the chauffeur. And she had three Rolls Royces.

We always went in the back door. You didn't think you were going up to that big iron front door, did you? Mercy on us! I remember one time I had two operations, and I wasn't allowed to come out, by doctor's orders, to do any cooking for a certain number of weeks. And Mrs. Worthingham, wealthy as she was, came up to me one day and told me, "Well now, you should be well enough to be able to work." I resented it, because I wasn't.

And I also resented the fact that Mrs. Worthingham came to me and said, "I do not want you to buy"—I did the ordering—"to buy any roast beef for the help. They don't need roast beef. You can buy other things, but not roast beef." And I said, "The people today have to have roast beef. It's a must, at least once a week, for the blood." Mrs. Worthingham and me had quite some words over it, but, you know, I won out, and I ordered that roast beef once a week for the help. If it was good enough for her, it should be good enough for the help that was doing the work for her.

Bridget Fitzgerald, who arrived in New York in 1921, found a job with a family so rich that even the servants had servants.

As soon as I landed, I went to an agency where they had special jobs.... They sent me to this job—it was a big estate out in the country and it was all for one woman. It was one of the great families. I shouldn't say the name. She had fifty people working for her—parlor maids, chefs, cooks, personal maids, two chauffeurs, two footmen in the pantry. There were so many of them...everything from all over the world. Lots of English. The butler was English. The footmen were English, but the housekeeper was Irish like me.

I was a "useful girl." You work one day a week in each department—with the butler, the parlor maid, the chamber maid, the personal maid, the footman, and you learn everything. It's good for you. You watch, you use your head. You pick it up—how to set a table, how to arrange fruit, how to arrange flowers, how to polish fruit, how to set a banquet table. Everything's got to be just so. You have to know how to set your wine glasses, your water glasses, different kinds of glasses. If you're going to serve fish, you got to know where your fish fork goes, where your salad fork goes, where your butter plate goes, your bread plate goes, your napkin goes. You have to make beds their way and dust their way. Lay out clothes, fix stockings....

I had my own bedroom, servants' quarters—a bathroom to each two girls. They'd give you a clean uniform every day. At night there were two men who took your uniform and the next day you had a clean one.... I wore a plain white uniform, all white—no cap. The parlor maid wore a cap, but I had my hair cut straight, bangs and straight; that was the style.

There were three cooks for the servants. Servants ate together in a large dining room. The food was out of this world. You'd get roast beef, chicken, fish, fresh corn. It was a long table with chairs and everybody helped themselves. And you cleaned your plate and put it back on the side, and they had help that took those plates away and washed them. You didn't do it, you know.

I'd have off from two to five and then I'd go back to work. And a half-day off in the whole week. That's all anybody got. You could go out between two and five, if you wanted to go out, but it was in the country. You could do whatever you wanted to do at that time—go to bed, go to church. There was a chauffeur to drive you to church if you wanted to. They supplied cars for you. You'd be surprised how nice those people can be to you. You would never know that they had money. They treat you like a human being. Never look down on you as though you're beneath them.

James Curley never forgot his mother's ordeal as a scrubwoman. When he became mayor of Boston, he took steps to improve their conditions. He wrote:

My mother was obliged to work...as a scrubwoman toiling nights in office buildings downtown. I thought of her one night while leaving City Hall during my first term as Mayor. I told the scrubwoman cleaning the corridors to get up: "The only time a woman should go down on her knees is when she is praying to Almighty God," I said. Next morning I ordered long-handled mops and issued an order that scrubwomen were never again to get down on their knees in City Hall.

Mary Harris "Mother" Jones at the celebration of her 100th birthday. She had become a legend among the union workers of the United States.

UNIONS

Irish workers were among the first union organizers in the 19th century. Their tradition of banding together to fight the English landlords served them well. Yet it was a long and hard struggle to form strong bonds with other workers and to wage a successful strike. David Lawlor described some of the difficulties in organizing the mill workers of New England.

I went through many strikes and lockouts which Fall River [Massachusetts] suffered for years and years. I never knew the employees to win a strike. The time seemed always selected by the manufacturers. Once a strike lasted twenty-six weeks. People had no money, they paid no rent, no bills, and as it was in the summer they lived on huckleberries and fished and dug clams, but during that long strike many of the old families were seen to go over the bridge with their little household effects.

Mary Harris Jones was born in Ireland and came to the United States as a child with her family. She married an iron molder and unionist in 1861 and became the mother of four children. In 1867 yellow fever swept through Memphis, where she lived, and Mother Jones's husband and children all died. She moved to Chicago and opened a dressmaking business. There she became aware of the great gulf between rich and poor and joined the labor movement.

Often while sewing for the lords and barons who lived in magnificent houses on the Lake Shore Drive, I would look out of the plate glass windows and see the poor, shivering wretches, jobless and hungry, walking along the frozen lake front. The contrast of their condition with that of the tropical comfort of the people for whom I sewed was painful to me. My employers seemed neither to notice nor care....

The Knights of Labor was the labor organization of those days. I used to spend my evenings at their meetings, listening to the splendid speakers....

I became more and more engrossed in the labor struggle and I decided to take an active part in the efforts of the working people to better the conditions under which they worked and lived. I became a member of the Knights of Labor.

Mother Jones dedicated her life to the labor movement. She appeared wherever workers were waging a strike. In her autobiography she described her experience with coal miners in Lattimer, Colorado.

Lattimer was an eyesore to the miners. It seemed as if no one could break into it. Twenty-six organizers and union men had been killed in that coal camp and in previous strikes. Some of them had been shot in the back. The blood of union men watered the highways. No one dared go in.

I said nothing about it but made up my mind that I was going there some night.... The general manager of Lattimer said that if I came in there I would go out a corpse. I made no reply but I set my plans and I did not consult an undertaker.

From three different camps in the Panther Creek I had a leader bring a group of strikers to a junction of the road that leads into Lattimer. There I met them with my army of women again.

As I was leaving the hotel the clerk said, "Mother, the reporters told me to ring their bell if I see you go out."

"Well, don't see me go out. Watch the front door carefully and I will go out the back door."

We marched through the night, reaching Lattimer just before dawn. The strikers hid themselves in the mines. The women took up their position on the door steps of the miners' shacks. When a miner stepped out of his house to go to work, the women started mopping the step, shouting, "No work today!"

Everybody came running out into the dirt streets. "God, it is the old mother and her army," they were saying.

The Lattimer miners and the mule drivers were afraid to quit work. They had been made cowards. They took the mules, lighted the lamps in their caps and started down the mines, not knowing that I had three thousand miners down below ground waiting for them and the mules.

"Those mules won't scab today," I said to the general manager who was cursing everybody. "They know it is going to be a holiday."

"Take those mules down!" shouted the general manager.

Mules and drivers and miners disappeared down into the earth. I kept the women singing patriotic songs so as to drown the noise of the men down in the mines.

Directly the mules came up to the surface without a driver, and we women cheered for the mules who were first to become good union citizens. They were followed by the miners who began running home. Those that didn't go up were sent up. Those that insisted on working and thus defeating their brothers were grabbed by the women and carried to their wives....

The victory of Lattimer gave new life to the whole anthracite district. It gave courage to the organization. Those brave women I shall never forget who caused those stone walls to fall by marching with tin pans and catcalls.

The Molly Maguires

The Irish, using a tactic they had employed to fight the British, sometimes formed secret labor organizations to win their battles for better pay and working conditions. The best known of these secret groups was the Molly Maguires, supposedly named after a woman who organized resistance against landlords in Ireland. (Indeed, some historians question whether the Molly Maguires ever existed as a formal organization.)

A long strike in the coal fields of Pennsylvania from 1874 to 1875 was said to have been led by the Mollies. The strikers used sabotage and violence to keep the mines closed, and were accused of murdering mine officials and police hired to guard the mines. An Irish-born detective, James McParlan, infiltrated the organization, and his testimony helped to convict 20 Molly Maguire leaders. Their executions brought an end to the organization, but the Mollies inspired the creation of more effective labor movements.

Luther "Yellowstone" Kelly (standing) won lasting fame as one of the mountain men who explored the West from the Dakota Territory to the Rocky Mountains.

IRISH IN THE WEST

Although the vast majority of the Irish immigrants settled in the cities of the East, the Irish played an important role and became successful in the West as well. The Irish arrived in the West even before it was part of the United States. The Texan Irish settlements of San Patricio and Refugio were founded in the 1830s, when the region was part of Mexico. Irish soldiers fought for Texan independence and served as Confederate soldiers from that state in the Civil War.

The Irish moved West first as trappers and guides. Later they worked for the railroads and participated in the mining booms throughout the region. Some Irish got very rich. The Comstock Lode, the richest vein of silver ever discovered, was owned by Irishmen. Marcus Daly, whose Anaconda Mines in Montana made him a millionaire, welcomed Irish workers to Butte. The city's Little Dublin was one of the largest Irish settlements in the West. In addition, the Irish were well represented among the soldiers and lawmen of the West. They also appeared on the other side of the law as well. Such outlaws as Billy the Kid (William Bonney) and Butch Cassidy are part of American folklore.

Irish-born Thomas McGuill went to Texas in 1853. Working as a peddler, he earned enough to bring over his wife and children from Ireland. Eventually, he opened a store and built a log church for the Catholic families in the area. The photo shows a Sunday gathering at the McGuill home.

Jeremiah O'Sullivan, of Virginia City, Nevada, wrote to his brother in Ireland on July 29, 1859.

You will all have been wondering what has become of us since we sailed from New York.... I shall only say here that our voyage from Ireland last year was as nothing when compared to this latest voyage.

I will get on at once to the great news. We were not long arrived in this place when Owen and I found the treasure we had come to seek, yes, my doubtful brother, we have indeed struck a vein of purest silver, and are in prospect of becoming very rich men. The settlement here is full of Irish, along with what seems to be at least one of every nation in the world. But we are among the few who have had the good fortune to "strike it rich" as they say here. It will be sufficient ore from our mine to repay our costs, but in a year's time, our fortunes will be made.

Now, Dan, why don't you come out here and join us? I know that you were not a mere scoffer, like those others who said Owen, Mary and I were fools to go to America at all and, bigger fools to head West. You were always one to stay home because you loved the land, and it will still hold you, I know. But you can make your fortune here too. It is not just a dream.

Rosalie Hart first heard about Texas when she was a child in Ireland. As she recalled many years later:

About this time in 1833, the Mexican government sent Colonel Powers to Ireland to get a colony of emigrants to settle in Texas. He belonged to a very good family, and had been raised in Exford. His only sister was married to a farmer named O'Brien.... Col. Powers held meetings at his sister's house, and made speeches to large assemblies. He represented Texas as one of the richest countries in the world, and the most delightful climate. Gold was so plentiful, according to his account, you could pick it up under the trees....

The doctors had always told father he would have to send me to a warmer climate if he wished to raise me, that I would certainly die in Ireland. The climate was too damp and cold for one as delicate as I was. Here was a good chance to save me.

The Hart family immigrated to Texas, where conditions were not quite what Colonel Powers had described. Rosalie Hart eventually became a schoolteacher to support her aging mother and her own two children. She takes up the story again:

The first two weeks that I was teaching school, there was neither corn [nor] flour in the country nearer than twenty miles. I succeeded in getting a few bushels of corn, but it was so badly eaten by weevils and so musty that it would make one sick to smell it, yet it was better than nothing. This I ground on a hand mill and sifted the best of the meal to make bread for my mother and my children. The hulls I carried to the schoolroom to make bread for myself; this and coffee

Death was a familiar presence among the homesteading families of the 19th century. The Breen family in Kansas buries one of its children but takes consolation in the belief that they would be reunited in heaven.

Molly Brown

Molly Brown was among the passengers who boarded the *Titanic* for its maiden voyage in April 1912. Its staterooms were filled with other wealthy people who wanted to be among the first to make the transatlantic crossing on one of the largest and most luxurious ships ever built. The mood aboard ship was festive, for the *Titanic* was said to be unsinkable because a new type of design had been used in its construction.

Molly Brown had come a long way to be able to afford this journey. She was born Margaret Tobin in 1867 in Hannibal, Missouri, the daughter of poor Irish immigrants. She worked in a tobacco factory and as a waitress before heading for the mining town of Leadville, Colorado. There she married James Brown, one of the miners who dreamed of striking it rich. Unlike so many others, Brown actually found gold. With his newfound wealth, he built a grand house in Denver. Molly (as she liked to be called) invited Denver's leading families to a ball to make her entrance into society. But they snubbed her, for even wealth could not overcome the prejudice against her "shanty Irish" background.

Molly Brown took her wounded feelings to Europe, where her outspoken manner and loud laugh were accepted as typically American. She married her younger sister to a German nobleman. Returning to the United States after one of her overseas trips, she booked passage on the ill-fated *Titanic*. When the ship struck an iceberg and began to sink, the lifeboats (too few to hold all the passengers) were offered to the women and children.

Molly Brown rallied the weeping occupants of her lifeboat, forcing them to take the oars and row. Though the *Titanic* had failed to live up to its billing, Molly declared that she was "unsinkable." After being rescued, the other women in the lifeboat told the story, and newspaper accounts made the "unsinkable Molly Brown" a heroine. Since her death in 1932, plays and movies have celebrated the career of this indomitable Irish American woman.

was the only food I had during those two weeks, except when one of the scholars brought me a piece of fresh meat. But this I carried home at night....

We were fortunate that year to find a family who proposed cultivating our field on shares. They were very industrious and did their work well, and although no one else with the exception our family on the Papalote raised anything on account of the drought, we had the best crop that was ever raised in that section of the country. We had, for our share, 300 bushels of corn besides a great quantity of melons, pumpkins, and other small vegetables, and if the corn had not been stolen while in the field we would have had a great deal more.

Mining work was dangerous, and tragic accidents often occurred. Catherine Hoy, the daughter of Irish immigrants, worked in the mining town of Butte, Montana, in the early 20th century. She told an interviewer:

I don't know if you remember the Granite Mountain disaster. There were, I think, forty, fifty men killed. They used this road to take the bodies back and forth. That was a memory because, you know, you figure so many families lost fathers or brothers or uncles or somebody. It pertained to somebody. Everybody in the vicinity suffered the loss of somebody like that.

They were a very close community. They never wanted for anything. Because if they did, there was somebody there right ready to help. I don't think there was any selfish people in that town. I lived there for twenty-five years. Up on that road. Lots of fond memories, too.

I was a bucket girl, what they called a bucket girl [serving meals to the miners at McManamon's boardinghouse, where she worked].... If the miners were nice to you, you gave them an extra cupcake or an extra piece of cake or an extra piece of fruit or somthing. But if they weren't, you know...I mean smart alecks or something like that, they just got their usual sandwich. And this was all made with homemade bread. The slices were about a half an inch thick. You slapped the meat in there and then you put another big slice on there. They really had to use their jaws to get around that. Some of those sandwiches....

The miners, you know, a lot of them didn't have homes, so they'd board and room in these boarding houses. They would have the big tables full of food all of the time. Food that you would never believe. Those were the men, single men, they had their bottles all the time.... There was a lot of rooming houses, too. I think they only paid about three dollars a week for a room. There was quite a few boarding and rooming houses there on Granite Street. The Big Ship, the Broadway, all those places. Dorothy Block. The Florence was another one. The Florence was on Broadway. The Big Ship was on Granite.

Many Irish who went west worked in the mining industry. This group was photographed at a mine in Rocky Bar, Idaho, around 1890. Rocky Bar mines had started producing gold in the 1860s.

Nora Clough (left) and her mother in the kitchen of their Arizona home around 1900. The Irish population of Arizona soared after the discovery of a huge deposit of copper in the early 1890s.

Minnie Finnegan (at right, in carriage) was chosen "Queen of the Fair" in Boise, Idaho, in September 1899. Her maids of honor were Rebecca Hays and Theresa O'Farrell. Lured by mining and logging jobs, many Irish immigrants settled in the West.

James Michael Curley was defeated in his attempt to win election to the Boston City Council in 1899. He noticed that many of the voters had simply marked an X next to the first name on the ballot. Curley was determined that next time his name would head the list. To accomplish this, he had to be the first candidate to file the nomination papers at the registrar's office. Curley arrived at the office the night before with some tough friends:

It was some siege. Several times that night flying wedges of rowdies tried to crash our lines, but we plugged them as they came in. John [his brother] suffered a broken jaw, and my cohorts and I took a pounding, but when the clerks and the registrars arrived the following morning, we still held the fort, and my name topped the ballot. Years later, the law was changed, and ballot position was determined by lot.

Boston's popular Mayor James Michael Curley and his son ride in the 1917 Evacuation Day Parade, which celebrates the day the British left the city during the Revolution. Curley, born in a Boston slum to immigrant parents, won election as congressman, mayor, and governor during his long political career.

POLITICS

Arriving in great numbers during the famine and afterward, the Irish immigrants used their votes to dominate the political machines in the big cities where they settled. The colorful Irish American mayor of Boston (and later governor of Massachusetts) James Michael Curley gave his view of why politics suited the Irish.

The Irish turned to politics because they had a natural affinity for it. When three Germans gather, they are apt to open a brewery. Put three Irishmen together, and they are likely to form a political club. The Irish turned to the political ladder also because it was the quickest and easiest way out of the cellar. The leaders helped themselves, but in so doing they helped their people, branded by the Brahmins [upper-class Protestant Bostonians] as "the shanty Irish." They became "lace-curtain," "cut-glass," "suburban," "Venetian blind" or the "F.I.F.'s" ("First Irish Families"). The step up the ladder could be swift indeed. Many of the Irish climbed by personal politics....

I remember a local character named Jimmy Walsh in whose behalf I appealed to Mayor Patrick A. Collins for a city job. His Honor said he was too busy to attend to such a detail, but changed his mind when I told him that Jimmy had voted for him twenty-eight times in the 1903 mayoralty contest.

One of the Irish American political leaders who "helped himself" was George Washington Plunkitt. Plunkitt worked for Tammany Hall, the New York City Democratic political machine, for 45 years. In 1905 he gave an interview in which he described how he was able to deliver votes for the machine.

What tells in holdin' your grip on your district is to go right down among the poor families and help them in the different ways they need help. I've got a regular system for this. If there's a fire in Ninth, Tenth, or Eleventh Avenue, for example, any hour of the day or night, I'm usually there with some of my election district captains as soon as the fire engines. If a family is burned out, I don't ask whether they are Republicans or Democrats, and I don't refer them to the Charity Organization Society, which would investigate their case in a month or two and decide they were worthy of help about the time they are dead from starvation. I just get quarters for them, buy clothes for them...and fix them up till they get things runnin' again. It's a philanthropy, but it's politics, too—mighty good politics. Who can tell how many votes one of these friends brings me? The poor are the most grateful people in the world.

Plunkitt also praised the Irish immigrants' achievement in politics.

One reason why the Irishman is more honest in politics than many Sons of the Revolution is that he is grateful to the country and the city that gave him protection and prosperity when he was driven by oppression from the Emerald Isle....

Yes, the Irishman is grateful. His one thought is to serve the city which gave him a home. He has this thought even before he lands in New York, for his friends here often have a good place in one of the city departments picked out for him while he is still in the old country. Is it any wonder he has a tender spot in his heart for old New York when he is on its salary list the mornin' after he lands?

In 1893 Finley Peter Dunne, a Chicago newspaperman, created the fictional Martin Dooley, a bartender and an Irishman from Roscommon. Mr. Dooley commented on everything from international affairs to his imaginary Irish American neighborhood, but his specialty was politics. He spoke in an Irish brogue that to modern eyes looks strange on the printed page, but Mr. Dooley's words remain as funny as they were when Dunne first wrote them. Here he described the duties of an alderman on the take named O'Brien ("O'Broyn").

No work or worry. Nawthin' but sit down with y'er hat cocked over ye'er oeye and' ye'er feet on a mahogany table an' let' th' roly-boly [easy money] dhrop into ye'er mit. Th' most wurruk an aldherman has to do is to presint himself with a gold star wanst a year so e won't forget he's an aldherman....'Tis good f'r annythin' fr'm a ball to a christenin' an'by gar Billy O'Broyn wurruked it on th' church. He went to mass over by Father Kelly's wan Sundah mornin' to square hisilf, an' whin Dinnis Nugent passed th' [collection] plate to him he showed th'star.

"Are ye an aldherman?" says Nugent. "I am that," says O'Broyn. "Thin," he says, stickin' th' plate under his nose, "thin," he says, "lave half f'r th' parish," he says.

Good-government reform groups (contemptuously called "goo-goos" by Mayor Curley) often tried to defeat the Irish political machines. Mr. Dooley also expressed a cynical view of the so-called reformers. He commented to his friend John McKenna:

This her wave iv rayform, Jawn, mind ye, that's sweepin' over th' counthry, mind ye, now, Jawn, is raisin' th' divvle, I see be the pa-apers. I've seen waves iv rayform before now, Jawn. Whin th' people iv this counthry gets wurruked up, there's no stoppin' thim. They'll not dhraw breath until ivery man that tuk a dollar iv a bribe is sint down th' r-road. Thim that takes two dollars goes on th' comity iv the wave iv rayform.

In 1920 a Chicago ward politician described the importance of the local saloon in political organizing:

Years ago, before what was called the regular organization, politicians would go to the corner saloon, and the saloon-keeper was the doctor, lawyer, banker—and that's where a candidate would go on Sunday to meet the people of the community....

Old timers would borrow money, come in for advice. The bank was at 63rd and Halsted. People wouldn't go there on the average—they'd borrow money from the saloon-keeper. A fellow running for office would visit ten or twenty of these spots. It was his way of campaigning!

Michael "Hinky Dink" Kenna (front left) and John Powers (right) at a political rally in Chicago around 1910. Though Irish-dominated political organizations were often criticized as corrupt, their leaders won support by caring for the needs of voters. In the recession of 1893, Kenna allowed people to sleep in Chicago's city hall and distributed meals to 8,000 people in a single week—making sure to register each one to vote as a Democrat.

Indeed, the reformers seldom made inroads on the loyalty most Irish Americans had for the Democratic party. Vinnie Caslan, whose family came from Ireland, described the importance of the Democratic party in New York City for his family.

When the Irish people got off the boat, we used to kid that they stamped 'em Irish, Democrat. If ya needed a job, ya went over and saw some Irishman. My father was a cop for one day, and he didn't like it. Then he went down to Tammany Hall. He never got to Charlie Murphy [the boss]—that was going to the pope—but he saw his intermediary, and they put him in sanitation. That's where he worked till he died. He always voted straight Democrat. If there were ten Chinese against ten Irishmen, ya'd still vote for the Chinese if they were Democrats.

Perhaps the most colorful Irish bosses were "Bathhouse John" Coughlin and Michael "Hinky Dink" Kenna, who presided over Chicago's First Ward in the first half of the 20th century. Justice Paul Douglas of Illinois remembered them.

The Bath" and "Hinky Dink" ruled politically in the vice-saturated First Ward for approximately the half century from 1890 to 1940. They were legendary characters. The Bath was the bumbling and none too bright extrovert, while Hinky Dink was the shrewd and laconic introvert. They indeed resembled characters in the old Laurel and Hardy movies. The Bath, with his gaily colored waistcoat, his inability to speak a coherent sentence, his penchant for owning racehorses which couldn't or wouldn't run, and his sponsorship of absurd songs such as "They Buried Her by the Side of the Drainage Canal" and "Dear Midnight of Love," is, in a sense, as irresistibly funny as the Keystone Cops.

But not so was his partner, Hinky Dink. "The Hink" knew the price of everything and, his opponents alleged, the value of nothing. It was charged that every saloon, every gambler, every prostitute had to come across at rates which were presumably fixed according to the ability to pay. Mike, with his aides, saw to it that this was done and that the vote was delivered on primary and election days.

I have always thought that Hinky Dink had the makings of a great idiomatic Latin Scholar, for in his saloon across the longest bar in the world he had blazoned in bold letters, *In Vino Veritas*. When asked what that meant, he replied, "It means that when a man's drunk he gives his right name." I have never heard any of my professorial colleagues do better than that.

Michael Donohue, who began to look for work during the Great Depression of the 1930s, recalled the helpful role of New York City's political clubs.

The politics then were dominated by the clubs. If you were poor—and there were so many poor people around—you had to depend on clubs to act in a humanitarian way. When kids were in trouble, people looked immediately to the clubs to try to get them out of it. If you called them, they'd send a young lawyer who was working for them, and looking to get ahead, and he'd show up in court. In retrospect, I think the clubs served humanity a damn sight more than many of these great charitable institutions today with their tax write-offs.

I had an aunt who lived on the West Side [of Manhattan]—she ran a boarding house around 68th Street—and she was very active in Democratic affairs. Her name was Mrs. Kelly. She would introduce me to people who she thought might be helpful in getting me jobs. I also used to play on the Amsterdam Democratic Club baseball team and they eased the way into job situations....

Then one of the Democratic clubs promised one of my brothers a job and he couldn't take it, and he gave the ticket to me. They approved of me, and I went to work for Con Edison [the local electric company].

Alfred E. Smith won election in 1918 as governor of New York State with the support of the Tammany Hall machine. Nonetheless, he was by many accounts one of the best governors the state has ever had. When Smith became the first Catholic to receive a major-party nomination for President in 1928, he faced anti-Irish and anti-Catholic prejudice. He defended himself in a letter to a critic:

In your letter to me in the April *Atlantic Monthly*, you "impute" to American Catholics views which, if held by them, would leave open to question the loyalty and devotion to this country and its Constitution of more than 20 million American Catholic citizens...

You imply that there is conflict between religious loyalty to the Catholic faith and patriotic loyalty to the United States. Everything that has happened to me during my long public career leads me to know that no such thing is true. I have taken an oath of office in this state nineteen times. Each time I swore to maintain and defend the Constitution of the United States.... I have never known any conflict between my official duties and my religious belief. No such conflict could exist. Certainly the people of this state recognize no such conflict. They have testified to my devotion to public duty by electing me to the highest office within their gift four times. Nevertheless, Smith lost the election, a defeat that convinced many Irish American Catholics that they would never see one of their own as President.

Tammany Hall in New York City was one of the most powerful of the Irish American political machines. Members of the Society of St. Tammany, as the organization was officially known, enjoyed social occasions. Here, at a fancy costume ball in 1906, the guests of honor are dressed as Uncle Sam and the Spirit of Liberty.

The Garrity family of Yonkers, New York, in 1942.

PUTTING DOWN ROOTS

Thomas Fitzgerald, the immigrant great-grand-father of President John F. Kennedy, told his son why he gave up his dream of starting a farm in America. Fitzgerald said that he remained in Boston, working in his brother's grocery store, because of his wife, his relatives, and the church.

Family, religion, and community mattered deeply to the Irish immigrants. In the face of prejudice and poverty, they drew strength from the close-knit communities in which they lived.

From about 1820, New York's Irish immigrants settled in the Five Points section of lower Manhattan, named for an intersection where five streets met at a place with the ironic name of Paradise Square. By 1840 it was known as the most dismal slum in America. Families squeezed into tiny rooms and damp cellars for rents as low as two dollars a month. Garbage was thrown into the street, where it nourished rats and pigs. "Hot corn girls" sold ears of corn from boiling pots of water, and barefoot children begged for coins or sold newspapers.

After 1851, when New York's elevated railroad was extended north to 30th Street and Tenth Avenue, Irish immigrants followed.

Many found jobs on the nearby docks, and their children swam in the Hudson River. But the air stank of fumes from nearby slaughterhouses and breweries, and only the tough survived the frequent street battles there. According to one story, a young policeman compared the neighborhood to hell. His more experienced partner commented, "Hell's a mild climate. This is Hell's Kitchen," and the nickname stuck.

Irish immigrants lived in similar surroundings in other cities. In New Orleans the Irish settled along a canal they had been hired to dig, and the neighborhood acquired the name Irish Channel. St. Patrick's Church, constructed in 1838 around an old shed where the first Irish immigrants attended mass, still stands today. Before the Civil War, the New Orleans Irish were hired for jobs considered too dangerous for black slaves. The slaves were valuable property, whereas the Irish could easily be replaced when the next boat arrived.

The North End of Boston, the neighborhood around the wharves of Boston Harbor, became the city's most densely populated area with the arrival of tens of thousands of Irish in the mid-19th century. There, as in every Irish American community, the first Irish-owned businesses always in-

cluded a saloon. It was a meeting place, a political headquarters, and a refuge from the misery that the Irish immigrants endured. Patrick Kennedy, President Kennedy's grandfather, was a saloon keeper, and so were many leaders of the early Irish American communities.

The word *whiskey* comes from a Gaelic word (*uisgebeatha,* or "water of life"), and Irish whiskey, with its distinctive taste of Irish peat, was as popular in the United States as in Ireland. Alcoholism, unfortunately, was a persistent problem in Irish American communities. Those who filled the saloons on Saturday nights were often chastised the next morning in a nearby Catholic church, where the topic of the priest's sermon was likely to be the evils of drink. Though temperance societies tried to combat this social problem, they never overcame the allure of the saloon as a place to drown one's sorrows with boon companions.

Alcoholism was one reason why a comparatively large number of Irish American families were headed by women. When husbands died young from disease, injury, or overwork, Irish American mothers took on the responsibility of providing for their children.

The Irish American family was often an extended one. Both in Ireland and the United States, a considerable number of both men and women remained single, marrying in their 30s and 40s if at all. They joined the households of their relatives.

In Ireland some unmarried sons and daughters turned to the church as a career. Irish priests and nuns frequently traveled to America to serve the growing numbers of Irish American Catholics. By 1900 more than half of all American Catholic priests and nuns were Irish-born or descendants of Irish immigrants. Better educated than most other immigrants, they provided counsel and leadership to Irish American communities.

The American Catholic church also established a system of elementary and secondary schools, as well as hospitals and orphanages, largely staffed by nuns. This huge undertaking required considerable financial sacrifice from the immigrants, but Irish Catholics had a deep-seated distrust of public schools, partly because in Ireland the state schools were part of the effort to eradicate Irish culture and Catholicism. In America the Catholic school system protected the immigrants' children from secular and Protestant influences.

American Catholics also founded many colleges and universities. The first was Georgetown University, founded in 1789 in Washington, D.C., by Irish American John Carroll, the first American Catholic bishop. Although French priests founded the University of Notre Dame in 1842, it later became closely identified with Irish Catholicism. As in many other American Catholic colleges, a large proportion of Notre Dame's teachers, both priests and laity, were of Irish descent. Today, after most colleges founded by Protestant denominations have lost their religious affiliation, American Catholic institutions of higher

Strongly patriotic, Irish Americans have fought for the United States in every war since the Revolution. These children in an Irish American neighborhood are celebrating the end of World War I in 1918.

learning have retained their close ties to the church.

The American Catholic school system provided the education that enabled the children of immigrants to rise higher than their parents had. Greater numbers found white-collar jobs, started their own businesses, and entered the professions. But doctors, lawyers, teachers, and successful merchants tended to leave the old Irish neighborhoods for middle- and upper-class areas. Those who moved up in society became known as "lace-curtain Irish," as opposed to the "shanty Irish" who remained in poverty.

Irish American social and benevolent associations have a long history. The Friendly Sons of St. Patrick, composed of both Catholics and Protestants, began in New York City in 1784. Irish residents of Philadelphia started a Hibernian Society in 1790, and three years later some of its members founded the Society for the Relief of Emigrants from Ireland. Countless other such groups emerged in the 19th century. Some, like the Sligo Young Men's Association, founded in New York in 1849, consisted of people who came from a specific county in Ireland. Others, such as the Ancient Order of Hibernians, were open to all Catholics of Irish descent.

Most of these groups attempted to alleviate the harsh conditions in which Irish Americans lived. Some organized efforts to encourage abstinence from alcohol and to fight the prejudice that Irish Americans endured. A great many kept their ties to the old country and supported the ongoing Irish struggle for independence from Great Britain.

When refugees from the failed Young Ireland rebellion of 1848 fled to the United States, Irish Americans greeted them as heroes. Ten years later, two former Young Ireland members founded the Fenian Brotherhood, dedicated to winning Ireland's independence. The Fenians attracted followers in most Irish American communities.

The movement was interrupted by the outbreak of the U.S. Civil

War. Ever since the Revolution, Irish Americans had found careers in the army, and Generals George McClellan and Philip Sheridan were among the notable leaders of the Union army. Companies of Irish Americans fought for both sides in the war. Distinguished for their bravery, they carried green banners that proclaimed their pride in Irish nationality.

After the Civil War ended, some Irish American veterans of the Confederate and Union armies took up the Fenian cause again. Collecting funds at mass rallies in Irish American neighborhoods, they organized an invasion force—but the target was Canada, not Ireland. The Fenians' plan was to capture Canada, still a British domain, and exchange it for Ireland's independence.

This scheme resulted in two Fenian invasions of the United States' northern neighbor. In the first, in 1866, the Fenians crossed the border from Buffalo, New York, and scored a temporary victory before retreating. Four years later, a second Fenian invasion force again moved into Canada but was swiftly defeated.

However, another secret Irish American group, the Clan na Gael, soon arose. It raised money to purchase guns and supplies to keep Ireland's fight for independence alive. Under such leaders as John Devoy and Diarmuid O'Donovan Rossa, the Clan na Gael also worked closely with Irish groups such as the Land League. When Ireland experienced another failure of the potato crop in 1877, the Land League organized relief efforts for the farmers. Charles Stewart Parnell, an Irish member of Parliament, came to the United States in 1880 to appeal for contributions to the league. While here,

In 1920, 70 percent of Irish Americans lived in the large cities of the Northeast. On Sundays and holidays, they could enjoy a day at the seashore. The grandparents of historian Loretto Dennis Szucs are among this group on a train at New York's Coney Island amusement park around 1910.

Parnell was invited to speak before the U.S. House of Representatives—a sign of Irish Americans' political clout.

When England went to war against Germany in 1914, Irish rebels, financed partly by the Clan na Gael, staged an uprising in Dublin on Easter Sunday, 1916. The revolt was crushed after a week, and most of its leaders were executed. But Eamon De Valera, born in New York, was spared because he was an American citizen.

While imprisoned, De Valera was elected head of the Sinn Fein (Ourselves Alone) party. In the na-tional elections of 1918, Sinn Fein won three-fourths of all the Irish seats in the British Parliament. However, the victors formed their own assembly, the Dail Eireann, and declared Ireland a republic with De Valera as president. The British responded by arresting the members of the Dail Eireann.

De Valera escaped and went to the United States, where he rallied Irish Americans to support his government in exile. Irish American dockworkers refused to unload British ships in U.S. ports, and contributions poured in from Irish American communities.

Bloody conflict raged in Ireland until December 1921. At that time a treaty was signed, creating the Irish Free State from 26 of the 32 Irish counties. The other six counties, all in Ulster where Protestants were a majority, remained part of Great Britain.

Though De Valera opposed this settlement, he finally became head of the Irish government in 1932.

Irish Americans take pride in the fact that the man who led the struggle for Irish independence was American born.

At every step along the path, the Irish Americans had played a role. They contributed money, provided volunteers, and used their growing political influence to support the cause. Even today, some Irish Americans support the ongoing guerrilla movement of the Irish Republican Army to "liberate" the six counties of Ulster, which remain part of Great Britain.

NEIGHBORHOODS

In 1873 future New York governor and Presidential candidate Alfred E. Smith was born near what is now the Manhattan side of the Brooklyn Bridge. He recalled his neighborhood with fondness.

When I was growing up everybody downtown knew his neighbors—not only people who were immediate neighbors but everybody in the neighborhood. Every new arrival in the family was hailed not by the family alone but by the whole neighborhood. Every funeral and every wake was attended by the whole neighborhood. Neighborly feelings extended to the exchange of silverware for events in the family that required some extraordinary celebration....

The families spent their summer evenings on the sidewalks, sitting on chairs and camp stools, and the policeman, as he came along on the beat...stopped at each stoop to bid everyone "Good-evening."

A year after Smith's birth, another Irish politician, James Michael Curley, was born in the Roxbury section of Boston. Curley's memories of what was unmistakably a slum remained clear 73 years later, when he wrote his autobiography, I'd Do It Again.

My brother John and I used to go down to the dump and forage for unburned pieces of anthracite [coal] and scraps of wood. The dump on Southampton Street was a former canal, a convenient receptacle for garbage, ashes and other refuse local residents threw into it. An excellent breeding place for rats, too, I remember. I still have a horror of rodents.

Life was grim on this corned-beef-and-cabbage riviera. I can remember tramping around barefooted, and after a heavy summer's rain, a familiar sight was a wagon sunk hub-deep in the mire. I remember the rotting fish shining in the sun on the mud flats off Northampton Street, the big wharf rats that scurried around, and the flies and mosquitoes unobstructed by house screens.

We played "Duck-on-the-rock," "Relievo" and other boyhood games; went fishing for "shiners" in the Muddy River over in the Fenway, or swam in the nude in the ship channel in South Bay or dove off the docks. Sometimes we walked to the Dover Street Bridge where there was a bathhouse. In season the gang played baseball and football on vacant lots, on the flats or right in the street when no "coppers" were around. During the winter there were coasting, skating and spirited snowball fights.... There were plenty of gang fights, especially with the

The children of Irish immigrants grew up in the stifling rooms of urban tenement houses. Inside, there was no privacy. This young woman, waiting for her laundry to dry on the line outside, steals the chance to have a conversation alone with her boyfriend.

Playing in the city streets, Irish American children made up games and provided their own entertainment in the 19th century. As Ed McGee recalled, "We liked to run and jump on the dead horses" that were left on the streets.

"toughies from Southie," or with other "wise guys" who came around, sometimes armed with rocks and slingshots. Ward Seventeen had its rough-and-tumble teen-age gangs who hung around the gashouse or in the square at the end of the street. "Cheezit, the cops!" still has a nostalgic ring to me.

A licorice stick or a candied apple was a treat, and a jelly roll or an ice-cream soda, which cost a nickel, was an event. Some of the more sophisticated lads smoked cigarettes made of dried leaves or sweet ferns. There was the carnival held once a year on Boston Common, where deer grazed when I was a boy, and I can recall as a boy playing around the Old Roxbury Canal in a square-end flat-bottomed punt which in its design resembled the old Irish curragh [a small boat].

Catherine Hoy grew up in Dublin Gulch, the Irish neighborhood of Butte, Montana, in the early 20th century.

There were, I'd say, fifty to sixty kids in that area. They all went to St. Mary's School. We had to cross the railroad tracks from our house to St. Mary's School. And believe me, it was snow. It was knee-deep getting back and forth to that school. But it was nice and we got along real well. Lots of fights and lots of quarrels. But they managed. There was one thing about kids on Anaconda Road and the people of Dublin Gulch: they all stuck together. They fought and quarreled, you know, had their fights and quarrels, but they stuck together. If one was in trouble, you'd best know, everybody was on the helping side of it....

As Others Saw Them

In 1831 a visitor described the Irish shantytown in Lowell, Massachusetts:

In the suburbs of Lowell, within a few rods of the canals, is a settlement, called by some, New Dublin, which occupies rather more than an acre of ground. It contains a population of not far from 500 Irish, who dwell in about 100 cabins, from 7 to 10 feet in height, built of slabs and rough boards; a fireplace made of stone, in one end, topped out with two or three flour barrels or lime casks. In a central situation, is the schoolhouse, built in the same style of the dwelling-houses, turfed up to the eaves with a window in one end, and small holes in two sides for the admission of air and light.... In this room are collected 150 children.

Mary Keegan (at left) owned this bakery in Lawrence, Massachusetts, photographed in 1916. Her friend Almeda King (right) wears a blue serge dress she made at the St. Clare School, founded by an Irish American organization to teach job skills to young people.

In 1893, William J. Kiley and two employees posed in front of Kiley's Market in Lawrence, Massachusetts. North of Boston, Lawrence attracted many Irish immigrants, who found work in its cloth mills.

From a very early age, children were expected to help support their families. In 1909, this group of happy youngsters scrounges through the Boston city dump for anything that could be sold or used.

The Overall Gang, they were a rough-and-tumble bunch of kids from age sixteen to about twenty, I imagine. Mischievous. They'd steal tires, hub caps, things like that. Or take a horse and buggy and put the horse in backwards. Or unharness the horse, and when the driver got in, the horse walked off and the wagon stood there. They got blamed for everything. I'm not saying they were angels, because they weren't. They'd go down in Finntown and all those places to fight. They didn't do it in their own community. Anaconda Road was on a hill and Finntown was down below us on Granite, you know, and we could pelt rocks down there, but they couldn't pelt the rocks up....

They used to call me "Fatty" at one time when I was a kid. But very few did without a black eye or something. All in all, it was a nice community. A nice place to live.

Michael Doyle described his home in San Francisco around 1850.

An old, unpainted clapboard house, without so much as a brick under it, set on the shifting slant of a sand hill...the ground unfenced, neither well nor water pipe, no sewer or street lamp for years to come...not a twig for fuel anywhere in sight—think of such a habitation for five and only one an adult!

Water was brought in long water-carts whose drivers carried it up the bank to a barrel at our back door in five gallon buckets at stated times, but as this did not suffice where so many clothes must be kept clean, I had two smaller buckets and a hoop and brought free water from a pump at the "Mile House," only a half block westward. Coal was not within reach, so we burned oak delivered in cord measure. It was my work to saw this up into stove lengths and split it as needed.

[A nearby pond] was the home of frogs and small turtles, i.e. terrapin. These last, in numbers would crawl up the western slope of sand to bask in the sun. Now, living near the place, I often stole down from the road, quietly waiting till a number were well out on the bank, then suddenly rushing knee-deep to head them off before the last reached the water. My bare shins might be chilled while waiting, but I did not mind when I was successful. Taking such prizes to school, I swapped them with boys for tops, knives and marbles. In these I was opulent. We boys, though, had no such chance to earn money as newsies and messengers have now. Only an occasional dime was gotten for some errand, gathering junk, or by catching large frogs for restaurants and such like. Oh, the mud our sleeves could acquire in the slime-probing pursuit—mother maddening!

Mary Thompson, a retired teacher, remembered growing up in the New York Irish neighborhood of Chelsea around the turn of the 20th century.

The low-rise buildings meant that everybody had access to the street, so that most of the children were watched not only by their own parents, but by everyone else's parents. As a matter of fact, when some of the old-timers get together, they often say, "Every time I did something wrong, I was beaten by three mothers." There was a lot of hanging out in the streets.

Ed McGee, born in the Hell's Kitchen section of New York in 1907, recalled:

We played marbles, Johnny on the pony, and cat and stick. That was where you had a stick and a small piece of wood. You hit the end of the wood with the stick, and when it went up in the air you hit it. You'd break a window, then run like a son of a gun.

They used horses for everything then. When the horse died, th'd put him out in the street for a day or two. We liked to run and jump on the dead horses. Once the bladder broke on 'im— oh boy, the smell. Quite a few of the kids used to make rings from the horses' tails. You would go back there and pluck the hair yourself, or, if you asked the driver, he'd do it for you.

During the 1930s, an interviewer spoke to an old Irishman who grew up in the area of Newark, New Jersey, known as the Ironbound neighborhood, shared by Irish and German immigrants.

We didn't have modern conveniences in the houses. Ferrey and Market streets were the only streets paved. No, there was a side street paved. The only lights they had were lampposts—gas. The lamplighter used to go around with matches. He had to climb each lamppost, before he got smart enough to use a long light. We used to go swimming in the Passaic River with our birthday suits. I wish I had those days over again. That's when they had shad in the Passaic. We only had to open a few buttons and dive in. Those were the good old days. We didn't realize it. Lots of time we'd be diving from lumber piles at Clark's lumber house, when a policeman would come along. We'd swim across the Passaic and sit on the other side till he went away. The old Down Neck, when the Irish and German were here, was the best place in New Jersey—no, I mean in the world. Where the steelworks are we'd have ice skating thirty-one nights straight with the moon shining every night, when I was a kid. In the Blizzard of Eighty-eight we walked on the snow and looked into the second-story windows in all the houses of Market Street. There were two horse cars stuck between Monroe and Adams streets on Market. After we got wet from the snow, we built a fire in the horsecars and dried ourselves, so we wouldn't get a tanning when we got home.

John L. Sullivan

In 1877, when John L. Sullivan was 19, he went to see a boxing exhibition in Boston. The professional fighter in the ring challenged anyone to fight. Sullivan stepped forward. Two minutes later, he had knocked his opponent senseless and turned to face the crowd. "My name's John L. Sullivan," he shouted, "and I can lick any man in the world!" For the next 15 years, that boast was the trademark of the best-known boxer in 19th-century America.

Soon after Sullivan's first ring victory, he went on tour, offering $50 (more than a year's wages for many in those days) to anyone who could stay in the ring with him for four rounds. Few collected, and his manager arranged a fight with the world champion, Paddy Ryan.

Fight rules of the time allowed kicking and wrestling, and the boxers did not wear gloves. Each round lasted until one of the fighters was knocked down. The bout ended only when one of the opponents could not stand up. It took John L. Sullivan only nine rounds to knock out Paddy Ryan.

Sullivan was a hero to many, but to none as much as the Irish Americans. Songs celebrated his growing reputation. The last championship fight fought bare-knuckle style matched Sullivan against Jake Kilrain in 1889. It took 75 rounds, and two hours and fifteen minutes of battling, for Sullivan to win the victory. Afterward, he popularized the "Marquis of Queensberry" rules that set a definite time for each round and required the fighters to wear gloves.

On September 7, 1892, John L. Sullivan, overweight and out of shape, faced James "Gentleman Jim" Corbett in New Orleans. Sullivan had gotten roaring drunk the night before and rode to the fight singing. When Corbett knocked him out in the 21st round, there were no cheers. The crowd was silent. Although Corbett, too, was Irish, he never took the place of the "Great John L." in the hearts of Irish Americans.

Sullivan never fought again. He stopped drinking and gave lectures on temperance. When he died on his farm in 1918, thousands of mourners lined the streets of Boston.

The elderly members of Irish American families received the respect and support of their children and grandchildren. This reunion in Gussettville, Texas, in 1912, brought together the descendants of immigrants who had come to the region since the 1830s.

FAMILY

In 1855 Thomas D'Arcy McGee, in A History of the Irish Settlers in North America, *commented on the effect emigration had on the traditional Irish family.*

In Ireland every son was "a boy" and every daughter "a girl" till he or she was married.... They considered themselves subject to their parents till they became parents themselves.... In America, in consequence of the newness of the soil, and the demands of enterprise, the boys were men at sixteen.... They all work for themselves, and pay their own board. They either live with the "boss," "governor," or "old man," or elsewhere, as they please. They may have respect,—they must have some natural deference for parents, but the abstract Irish reverence for old age is not yet naturalized in America.

Margaret Higgins Sanger, who was to become the leading advocate of birth control in the early decades of the 20th century, described her father with affection.

Born in Ireland, Michael Hennessy Higgins was a nonconformist through and through. All other men had beards or mustaches—not he. His bright red mane, worn much too long according to the family, swept back from his massive brow; he would not clip it short as most fathers

"Proper" young Irish Americans did their courting under the eyes of their families. Around 1905, this group of young men and women gathered on a front porch in Massachusetts.

did. Actually it suited his finely modeled head. He was nearly six feet tall and hard-muscled; his deep blue eyes were set off by pinkish, freckled skin. Homily and humor rippled uneasily from his generous mouth in a brogue which he never lost....

The scar on my father's forehead was his badge of war service. When Lincoln had called for volunteers against the rebellious South, he had taken his only possessions, a gold watch inherited from his grandfather and his own father's legacy of three hundred dollars, and run away from his home in Canada to enlist. But he had been told he was not old enough, and was obliged to wait impatiently a year and a half until, on his fifteenth birthday, he had joined the Twelfth New York Volunteer Cavalry as a drummer boy.

Immediately upon leaving the Army, father had studied anatomy, medicine and phrenology, but these had been merely for perfecting his skill in modeling. He made his living by chiseling angels and saints out of huge blocks of white marble or gray granite for tombstones in cemeteries. He was a philosopher, a rebel and an artist, none of which was calculated to produce wealth. Our existence was like that of any artist's family—chickens today and feathers tomorrow.

Christmases were on the poverty line. If any of us needed a new winter coat or pair of overshoes, these constituted our presents. I was the youngest of six, but after me others kept coming until we were eleven. Our dolls were babies—living, wriggling bodies to bathe and dress instead of lifeless faces that never cried or slept. A pine beside the door was our Christmas tree. Father liked us to use natural things and we had to rely upon ingenuity rather than the village stores, so we decorated it with white popcorn and red cranberries which we strung ourselves. Our most valuable gift was that of imagination.

Mary Ann Kelly described how her grandfather Redmond, the son of Irish immigrants, started his own family in the 1870s.

Grandfather went to work for the railroad, following in the footsteps of other Irish immigrants. Soon afterward he met our grandmother Annie McHugh, [a] young Irish servant girl in Cincinnati. "He had a yearnin' for her" as they put it in Ireland, so they got married and settled in Hinton, Kentucky.... Here in a little Southern Railroad house, painted gray like the depots, near the tracks, their eight children were born, four of whom died in infancy or early childhood. Our father and his two older brothers could remember the lonely burials of their little brothers and only sister in the small cemetery.

James T. Farrell, author of many novels about Irish American life, remembered his Irish-born grandmother:

She dominated the household, she was a spirit, she had the fear of God but I doubt she had any other fear in her. She and her older sister had "come out" from the Irish midlands.... The world never defeated them.

In close-knit Irish American families, house parties like this one in Wisconsin around 1910 included young and old in the festivity.

The children of Raymond Francis and Margaret Howley Dyer in the yard of their home in Brooklyn, New York, in 1911. From left, Margaret, Madelon, Muriel, Edwin, and Ethel Dyer.

Several generations contributed their work to the family business, like this store in an Irish American neighborhood known as Hell's Kitchen in New York City around 1890.

With the father often at work or absent, the Irish American mother frequently played a dominant role in family life. David Lawlor, who arrived in the United States as a small boy in 1872, described the importance of his mother.

My great companion was my mother. She sat up every night until her seven boys were in. They would come in, salute her and go up to their rooms. But I would always sit by the side of the fire and discuss the events of the day. I was seldom out after nine o'clock except when I went to night school and then it would not be much later than nine-thirty. The only heat in the house was in the kitchen and all of us boys in winter would take off our hats and coats and shoes in the kitchen and go up to the "refrigerator." It surely was cold in that attic, but we slept three in a bed so that we could keep one another warm. I got in the habit of saying my prayers in the kitchen and a thousand times or more my good mother would take me off my knees when I was fast asleep and send me up to bed. After I was married I used to go to see her once a week and sit and chat with her about the way the world was going.

Mother's room was off the kitchen and she had a feather bed which was brought from Ireland. The rest of us had straw beds, but anyone who got sick would go down to mother's bed and father would go to the attic. This gave her a chance to nurse the patient. My brother John had typhoid fever, and day and night he was nursed for seven weeks. The only boyhood illness I ever remember was when I was twenty-four. I had the measles—that was the night the Fall River City Hall burned down. I could see the flames but they would not let me leave the bed. When mother was sick unto death, we had no money for nurses so the family conclave decided that the girls would nurse during the day and the boys at night.... When my night came, I was there at the time agreed, shortly after nine o'clock. My sisters went to bed and I sat on the sofa in the parlor opposite the bedroom so I could watch the patient and answer her calls. I had worked hard all day and then I had taught school, and I was pretty tired, but I sat there as faithfully as any sentinel in the army. Nature asserted itself and I fell asleep, leaving my mother untended. When I woke after a deep sleep I found my head in my mother's lap. She found me asleep, got out of bed, put my head in her lap and she and my guardian angel watched over me. She died a few days after, and seven of us boys marched after the hearse to the cemetery. At the time it was a custom for the bearers to fill the grave. When the stones from the first shovel struck my mother's coffin, I thought it was the most terrible thing that had come to me in my whole life.

Catherine Hoy's family was an extended one with many members, but her most vivid memories were of her grandmother.

I came from a family of six, and each of us had our chores to do. We had six rooms. My older sister had the two front rooms. That consisted of a bedroom and a front room. My other sister had the kitchen, which was the hardest one of all. I had the dining room, we had a great big long table. It was always set, and on the table we ate our meals. You know, we'd sit around that. At one time, we'd have from ten to twelve at that table. We always had a cousin or an uncle or an aunt or somebody living with us, you know. Just the old Irish tie. We always had someone living with us. That area, that was mine. I had to do that. Every Saturday, whether it needed it or not, we had to scrub the walls down. We had to do this, and we didn't slipshod it. We used soap and water. Naptha soap and water. The table had to be scrupulously clean, or you didn't get your supper.

I had an aged grandmother who lived with us. She used to sit in a corner, a little corner that was off by the big stove, you know. She smoked a corncob pipe. Sometimes she couldn't light that corncob pipe so us kids took it on ourselves and we'd light the corncob pipe for her. That was typical old home and just one of those things that you'd just love to remember and cherish. The thought of it....

As I said, this aged grandmother of mine, she would gather most of the kids from around the neighborhood. They'd come in, and she'd tell them ghost stories all about the banshees and the Little People and all that, you know. Sh'd sit there for hours and tell us all those stories. Then the kids were too scared to go home, so my mother and my older brother would have to take the kids home. She lived in Ireland in those days when that kind of stuff was really true to them.

The Irish played an important role in American popular culture from the middle of the 19th century. There were many families—the Powers, the Rooneys, the Foys—who established show business dynasties. The comic genius Buster Keaton, a star of silent films, described his early days in show business with his family.

Having no baby-sitter, my mother parked me in the till of a wardrobe trunk while she worked on the stage with Pop. According to him, the moment I could crawl I headed for the footlights. "And when Buster learned to walk," he always proudly explained to all who were interested and many who weren't, "there was no holding him. He would jump up and down in the wings, make plenty of noise, and get in everyone's way. It seemed easier to let him come out with us on the stage where we could keep an eye on him...."

[Keaton eventually won a place for himself in the act.] This was the result of a series of interesting experiments Pop made with me. He began these by carrying me out on the stage and dropping me on the floor. Next he started wiping up the floor with me. When I gave no sign of minding this he began throwing me through the scenery, out into the wings, and dropping me down on the bass drum in the orchestra pit.

The people out front were amazed because I did not cry. There was nothing mysterious about this. I did not cry because I wasn't hurt. All little boys like to be roughhoused by their fa-

In 1902, Mary Judge (far right) posed with her daughter and grandchildren. Mary's husband, John, was the brother-in-law and business partner of Thomas Kearns, an Irish American who made a fortune in mining and was elected U.S. senator from Utah in 1896.

The Catholic Guardian, *a San Francisco newspaper whose readership was primarily Irish, gave the following advice to wives in the 1870s:*

The wise wife should always bear in mind that although the husband is the oak of the family, he is largely sustained by the little tendril of conjugal sympathy, encouragement, and praise; by these he is made brave and strong and persistent in the prosecution of his business, whether it be the tinkering of a tin pan, the mending of a shoe, or the building of a steamship.

The famine immigrants were known as "shanty Irish" because they sometimes lived in shacks. But some of their descendants became "lace curtain" Irish like this prosperous family in Boston.

thers. They are also natural tumblers and acrobats. Because I was also a born hambone, I ignored any bumps or bruises I may have got at first on hearing audiences gasp, laugh, and applaud....

Before I was much bigger than a gumdrop I was being featured in our act, The Three Keatons, as "The Human Mop." One of the first things I noticed was that whenever I smiled or let the audience suspect how much I was enjoying myself they didn't seem to laugh as much as usual.... [He developed the deadpan expression that made him a famous comic in the movies.]

Because of the way I looked on the stage and screen the public naturally assumed that I felt hopeless and unloved in my personal life. Nothing could be farther from the fact....

My parents were my first bit of great luck. I cannot recall one argument that they had about money or anything else when I was growing up. Yet both were rugged individualists. I was their partner, however, as well as their child. From the time I was ten both they and the other actors on the bill treated me not as a little boy, but as an adult and a full-fledged performer. Isn't that what most children want: to be accepted, to be allowed to share in their parents' concerns and problems?... It seems to me that I enjoyed both the freedom and privileges of childhood, certainly most of them, and also the thrill of being treated as full-grown years before other boys and girls.

In the United States, women had greater freedom in picking their husbands. Interviewed in the 1930s when she was 72, Marie Haggerty described her courtship and marriage years earlier. At that time, she was working in a household in Boston.

I was living with those folks when I met Pa. He was the grocery salesman and come for orders three times a week. I can hear him even now, for he was a great whistler, and very jolly. I could hear him a mile off, and I usually went out near the back, never letting him know, of course, but I always managed to make him see me, and he'd come and talk. He rode a horse and buggy, and they didn't deliver mail them days, so I would usually be on my way to the post office and he would drive me there and back. After a while, we got to keeping company, and we used to drive around the Cape Sundays.

I knowed Pa for three summers before we got engaged, and I well remember that day. It was a Sunday afternoon, and he come by with his horse and buggy. It was a hot day, so he tied up the horse, and we went walking. We walked down by the water, and he was very quiet, and there was people all around, so he said, "Kitten"—he always called me Kitten—"let's walk through this little woods, tairn't very thick." Well, I felt something was about to come, and I didn't know what. We walked for an hour or more, and then we set down on a tree stump, and while I was just picking grass and chewing on it, he outs with it, and asks me to get married. Mind you, first I was glad and said I would, but next I got mad. Tweren't like any proposal I'd ever heard of. I always thought when I was asked to be married, he'd do it kind of grand like—get down on his knees maybe. Wasn't I the fool? Well, I gave him one look and I ran away from him right down to the water where all the people was. He came after me, and then I got to feeling how silly I was, so I told him I wanted to go home and I went right into the house without even saying good-bye.

After I left I got to thinking about it, and I got sick to me stomach, for I had just about made up my mind never to get married but to learn to be a real nurse. When I got to my room and quieted down I decided definitely I wouldn't marry him at all. But the next time I saw him, he started to tease me about acting so and I couldn't help but feeling sorry for him, so I told him I didn't mean to act so. So that was the end of it. I married him.

We went to Boston to be married, for we was only summer people at the Cape. The lady I worked for let us have the coachman and the best carriage to go in, and when we got back to the Cape that night, they had a big party. It wasn't exactly like the rich people, but nearly. She had the gardeners and coachmen clear the barn for dancing, and the cook made up all the refreshments, and she gave us all the punch we could drink. Then before Mr. and Mrs. went to bed, they came out and drank to our health, and wished us their blessings and happiness. The only difference in my wedding and the rich people was that our party was in the barn; but it was nice there, and we had an accordion and a fiddler for music. All night long, as long as the party lasted, people came from all over with tins and pans and beat a serenade, and yelled for the bride and groom.

The Neely family of Florence, Oregon, enjoys a musical evening. The woman at right is popping corn over the fire.

Eugene O'Neill

Eugene Gladstone O'Neill was born in New York City in 1888, the son of James O'Neill, an immigrant from County Kilkenny. His father regaled the young boy with the glories of Irish history, particularly stories of the O'Neill clan. James O'Neill was an actor who had become a hit in the leading role of *The Count of Monte Cristo* and played the character more than 6,000 times. But his wife disapproved of her husband's profession and took refuge in drugs. James himself felt he had wasted his talents in the role. Eugene fled home at an early age to go to sea. But the theater was in his blood, and he started writing at 24 while recuperating from tuberculosis.

In 1916 a theater troupe called the Provincetown Players produced Eugene O'Neill's first play, *Bound East for Cardiff.* Its success brought him the opportunity to present his next work on Broadway. This drama, *Beyond the Horizon,* won him the Pulitzer Prize, the first of four he would earn in his lifetime.

In the next two decades, O'Neill was both prolific and successful. His plays dealt with topics that did not usually appear in the American theater. In *All God's Children Got Wings,* he treated the injustice of racial discrimination. He wrote about mistreated and abused workers in *The Hairy Ape.* But his most memorable characters were people whose dreams had gone sour, like those in *The Iceman Cometh.* In 1936 he won the Nobel Prize for Literature.

For much of the ten years before O'Neill's death in 1953, illness prevented him from working. But three years after his death, a new play appeared, one that he had written in 1940 but had not released because its subject matter was too personal. *A Long Day's Journey into Night,* his masterpiece, tells the story of the Tyrone family and was based on O'Neill's own family. He had come to grips with his tortured childhood by turning it into one of the greatest plays ever written by an American.

Sister Mary Dower and her father, Bill Dower, in New York in the 1940s. Catholic Irish American families contributed many of their daughters and sons to the church as nuns and priests.

CHURCH AND SCHOOL

Ninety-one-year-old Catherine Moran McNamara, who came to the United States in 1903 at the age of 18, recalled how she passed on her Catholic religion to her six children.

I went to my own Church, Catholic Church. We were all Catholics. They went to Saint Joseph's School, my children. And the boys, they had the Brothers then teachin' them, which was great teachers.

The Irish is brought up as good Catholics, good God-fearin' people. No matter how poor they were, they went on their knees every night and say the rosary, I remember that. We *never* seen a night when the rosary wasn't said in our house.

They had great faith, and they held the faith. Just like the Irish have suffered for the faith. Terrible too. The English wanted them all to abandon it. Did you know they used to burn their churches? If they'd be caught goin' to churches, they'd be jailed and shot dead one at a time. Imagine what my father seen when he was small. Imagine. He wouldn't describe it.

John Walsh, a longtime resident of Dorchester, an Irish neighborhood in Boston, was 82 when he was interviewed by a group of students from St. Gregory's School in 1978.

Q: *Did you ever go to St. Gregory's Sunday School?*

WALSH: Well, I was baptized in St. Gregory's. I can remember First Communion. I went to Mass on Sundays. You had to. They put the fear of God in you. I remember Father Heaney...me and another boy assisted at Mass. Then it was all in Latin. Well, Father Heaney didn't like the way we pronounced our words and kept us after Mass to say it for him. Father Heaney would hit you in a minute. He'd go out into the neighborhood and if he saw anything wrong he'd hit a kid in the street...and if he saw a girl out [alone] he'd tell her to get home. Even a Protestant girl, he'd tell her too....

Q: *What about the church fire at St. Gregory's; I think it was in 1871? I heard it was lightning.*

WALSH: Well, the Protestants burned it down. And when [the Catholics] tried to rebuild it they were threatened and told not to work on the church. I was told by my father it was blown up by the Protestants.

Q: *What about other priests?*

WALSH: Well, I remember Father Deegan in Sunday School talking about the Protestant church. He said, "The only thing right about the Protestant Church is the clock and that is wrong half the time."

Father McNulty used to say the 11:30 Mass and it was packed. His sermons were something. He'd tell them what Hell was. Everyone wanted to hear him. He later became the Pastor.

Almeda King was born in Lawrence, Massachusetts, in 1900. Both of her parents were immigrants. She graduated from Simmons College and became a teacher. In her nineties, she remembered the force of religion in her Irish neighborhood.

The community where we lived, South Lawrence, was almost all Irish—only two to four families were not Irish. Life in an Irish community was really worth having. I don't think you can give a better heritage than Irish. Everyone helped everyone else. They might have been nosy but everyone helped each other. The home and the Church were the great centers. You were born in your home and died in your home. The family asked the priest for guidance and the morals of the Church were carried out in the home. We had a good life. We knew to whom we belonged. We were close to the relatives— aunts and uncles visited all the time. Weekends especially were for family. The priests knew us all and we were friendly but formal with them. We never called the priests anything but Father Farrell or Father Smith. We felt proud of belonging to the Church and felt sorry for those who didn't.

At age seven, Catholic girls and boys received their First Holy Communion. In the 1950s, this group in the Sacred Heart parish of Boston poses with their pastor. Reflecting the pure state of their souls, girls wore white dresses and veils, and the boys wore white suits.

A class picture at Holy Innocents School in Brooklyn, New York, in 1917. From the mid-19th century until about 1950, the majority of priests and nuns in the Catholic church of the United States were Irish Americans.

The nuns who taught in Catholic schools received no pay; the parish provided housing, food, and clothing. Though a single nun taught a class that could include 50 students, she somehow managed to give attention to the needs of each one.

Every child was invited to the St. Pat's party at the parochial school. You could meet boys but the priests were always on hand. We learned Irish dancing, got ice cream and green candies—everything was green. I grew up thinking that St. Patrick was in Dublin.

Jane Byrne, who served as mayor of Chicago from 1979 to 1983, recalled her family's celebration of Holy Week—the week before Easter—in 1941.

My two brothers and I always looked forward to Holy Thursday Mass, because it marked the end of Lent and the daily rising at the grim hour of six to get to seven-o'clock Mass. It also signaled the fading away of cold dark winter and the first hints of spring, as tulips popped up beneath the evergreens beside our front door in Sauganash, one of Chicago's oldest neighborhoods....

This Holy Thursday was particularly special to me, because my oldest brother Billy, then twelve, had been chosen to lead the procession of parish schoolchildren, carrying high the cross that symbolizes Holy Week. Seven-year-old Edward and I marched proudly alongside our schoolmates, each boy paired with a girl in the same class. Dressed in white dresses and matching veils, all of the girls carried white lilies to lay upon the altar at Queen of All Saints Basilica, only six blocks from our house. There, in a ceremony that was both beautiful and filled with promise in my eyes, Billy assisted our pastor, Monsignor Francis J. Dolan, in serving the Mass.

When we reached home afterward, my two brothers and I were greeted by the inviting fragrance of warm hot cross buns and the sweet perfume of bouquets of Easter lilies. We couldn't wait to start dyeing our Easter eggs. Once they dried, we'd place them carefully in a rainbow of colors around the lamb-shaped cake my mother had baked and set out on a pedestal cake stand in the center of our dining room table. Our whole family loved Easter, but particularly my father, an unusually devout Irish Catholic even in a city famed for the breed.

Thomas Byrnes, an Irish American journalist, describes the importance of First Communion.

Each year in May, children in Catholic schools honored Mary, the mother of Jesus, with a Queen of May procession. One girl was chosen to lead the other students into the church, like this one in an Irish American parish in Massachusetts in 1916.

When I was a youngster, the occasion that drew the most members of my family together at one time was a First Communion. More so than a religious feast or a national holiday, a seven-year-old's introduction to the mystery of the Eucharist called for a total mustering of the clan.

Moreover, at no other time were you so exclusively the object of so much loving attention, except, perhaps, the day you were born.... And, you felt, deservedly so. Consider what you had just been through.

There was the period of intense pre-Communion instruction in the second-grade boys' class of Sister Mary Judith. Sister had been preparing boys for their First Communion for many years. To us kids she seemed quite old. Her First Communicants by now must seem quite old, too, but I'm sure that what she taught them is still as fresh in their minds as it is in mine.

It had to be fresh in *everyone's* mind the Friday afternoon before First Communion Sunday when Father Shay, our assistant pastor, conducted the oral examination. He sat at a card table just outside the open door of the classroom. One by one, thirty boys went out to be questioned, each convinced that unless he got the answers right he might *never* reach the altar rail.

"Now then," said Father Shay, his round face glowing with encouragement, "let's see how well you know the Act of Contrition. 'Oh, my God—'" he prompted....

"Corpus Domini nostri Jesu Christi custodiat animam tuam in vital aeternam. Amen."

You heard the words coming closer and closer as the priest moved down the altar rail from kneeling child to kneeling child. You mustn't turn your head to see how near he is. You must wait.

"Corpus Domini nostri...."

Suddenly the *corpus* was being offered directly to *you*. You tilted your head back. The last thing you saw before closing your eyes was the host tracing a sign of the cross above the chalice. You opened your mouth, and that strange new bread—that taste you had never known but would know forever—was resting on your tongue as gently as a prayer.

You marched back to your pew, eyes fixed firmly on the shoes ahead of you. But something was wrong. Your knees went rubbery as you realized that the host was stuck to the roof of your mouth, and you couldn't work it loose with your tongue. You were about to do the unthinkable—free it with a finger—when you recalled Sister Mary Judith's words, "If the host should cling to the roof of your mouth, don't worry. Just let it dissolve."...

Once home, you were promptly fed, of course. Before you dug in, you treated yourself to a moment of self-satisfaction at having survived the fast....

The rooms soon filled with the cheery sounds of arriving relatives, all bearing congratulations and gifts—so many gifts that you wondered what you would do with them all: seven prayer books, five rosaries, three statuettes of the Blessed Mother, two framed pictures of the Sacred Heart, innumerable scapulars, holy cards, and medals.

In time, some of them would disappear, as if by themselves. But others would work their way from the front to the back of dresser drawers to be come upon later, reminding you as you grew older that even on that special day, which you once thought was yours alone, you had shared the honors with someone else.

In 1890 the archdiocese of Boston issued the following statement:

This is the command of the Catholic Church, and she must be obeyed to please her Divine Father. Our Holy Father...says that these Catholic schools are the only places for Catholic children to receive their education and that a father or mother, claiming to be a Catholic, is a poor misguided Catholic who will send a Catholic boy or girl to schools where they may be certain, if they do not lose their faith, that they will, at least, have it weakened.

Two Irish American families gather at St. Mary's Church in Lawrence, Massachusetts, to celebrate a wedding.

Looking back, I see that it was religion that saved me. Our ugly church and parochial school provided me with my only aesthetic outlet, in the words of the Mass and the litanies and the old Latin hymns, in the Easter lilies around the altar, rosaries, ornamental prayer books, votive lamps, holy cards stamped in gold and decorated with flower wreaths and a saint's picture. This side of Catholicism, much of it cheapened and debased by mass production, was for me, nevertheless, the equivalent of Gothic cathedrals and illuminated manuscripts and mystery plays. I threw myself into it with ardor.

Besides schools and hospitals, Catholic religious orders also established orphanages, like the Kearns St. Ann's Orphanage in Salt Lake City, shown here. These nuns' distinctive style of wimple—the covering for their heads—indicates they were members of the Sisters of the Holy Cross.

In 1876 William O'Connell, born and raised in Lowell, Massachusetts, resolved to become a priest. It was the start of a career that would take him to the post of archbishop of Boston and to the rank of cardinal of the church. In the 1930s he recalled the reasons for that choice.

Our family life was an intensely religious one. As a youth I served at the altar. I loved the Mass with all its holy mysteries and beautiful decorous ceremonies, and I never felt so much at home as I did in the quiet church in the late afternoon, as I entered and rested in that atmosphere of peace on my way back from school to home. And as early in life as my fifteenth and sixteenth years, the thought of the sublimity of this service of God first awed me, and then by a gentle attraction, which seemed like nothing else but a mother's love, I began to wish and pray that I might be worthy of a call which could come from God alone. By the end of my high-school years in 1876...after a long conversation with my confessor...I finally decided to enter Saint Charles's College in Maryland [a seminary, or training school for the priesthood].

The Irish wake, in which friends and relatives gathered around to view the body of the dead person, was very much a social occasion, as Mary Devlin recalled.

Wakes were a great way to socialize. In those days, the people were waked in the homes for three days, and all the people would bring in food. The *Evening Journal* used to have big obituary columns, and my girlfriends would say, "Look up, see if you know anybody. It might be a good wake to go to." Instead of going to a party, we went to a wake. Half the time, the fellas would have the same idea, so you went there and you said your little prayer and paid your respects. Then you went to the dining room and had some refreshments. Nine times out of ten you'd meet somebody who'd say, "I'll walk you home." That was our social life.

Vinnie Caslan described his Catholic upbringing in New York City in the first half of the 20th century.

I went to church every Sunday till I grew up. It was mandatory. I went to school at Immaculate Conception. I had Christian Brothers. I think before they came in they had to be in the Golden Gloves.

I got a workout one time with the brother 'cause I was...doin' something I shouldn't have been doin'. So I come home with a black face and this and that all scuffed up. I didn't tell my mother, but she really gave it to me. I just said, "Aah, one of the guineas started trouble up there."

Anyway, on Sunday, my mother hits bing, bing, bing. She had been to church, and it turned out the brother who worked me over came from the same part of Ireland as she did. He meets her in church. "Oh, hello, Mrs. Caslan. [In brogue:] Did yer son tell ye about th' wallees I gave 'im? Oh, brother, I gave 'im quite a few wallees." So I got another workout from my mother. There's no such thing that he couldda been wrong. So I got the doubleheader.

Owen McGivern, born in 1912 in Hell's Kitchen, went on to become a judge of the New York State Supreme Court. He remembered the pastor of his church.

I went to the Holy Cross parochial school on 43rd Street between Eighth and Ninth avenues. The school was run by the Christian Brothers, and they ran it with a rod, no nonsense. If you didn't behave, they sent you to public school, so you behaved.

In 1920, Father Duffy became pastor of the church. I think he took an interest in me when I graduated from grammar school with some kind of medal. I got a scholarship to St. Regis High, and at that juncture, he became the biggest influence in my life. I saw a great deal of him then. He was a real scholar, and he had a vast library. We didn't have books around the house, so he would load up my arms with them....

[This was Father Francis P. Duffy, the chaplain of the "Fighting 69th" Irish regiment during World War I. Many young Irish Americans from Hell's Kitchen had served in the 69th.]

He was a very dominant figure, very tall, with a commanding presence. The public side of life didn't mean anything to him. He was well liked by the theatrical people, but that was because Broadway and the theaters dominated the area. Naturally, he knew all the actors and actresses. Spencer Tracy, [John] Barrymore, Walter Hampton—they were all friends of his. Many of them lived in the neighborhood.

He was the last guy in the world you would think to be a chaplain of a tough regiment, but he was deeply admired by his men. He was very understanding of human weaknesses. That's also why he was popular in the theatrical world. Coupled with his wide scholarly interests in politics, history, and literature, he was a very unusual man. Remember, many of the priests in those days were certainly not scholars. They were the sons of firemen and policemen whose mothers wanted to have a priest in the family. They went to a seminary, picked up Latin and learned how to perform the rituals of the Mass, but very few of them became great thinkers or theologians. Yet on Friday night, you'd see him, [deparment store owner] Bernard Gimbel, and [boxer] Gene Tunney walking up 42nd and up Eighth to go to Madison Square Garden to see the fights. He was a great fight fan.

He was the pastor of a large and poor parish. There was nothing high-hat about him. He'd walk up and down Tenth Avenue a great deal, and everybody knew him....

He was in many ways a great man. How many great men do you meet in life with unusual dominant personalities? I've met very few. You rarely meet the man with leadership, who has that magical "x" quality. He had that elusive quality.

Customarily, a "death card" was distributed at the wake or funeral of a deceased Catholic. It was kept as a memento of the person.

Carnivals, raffles, and bingo games were (and still are) popular ways to raise funds to support the parish and parochial school.

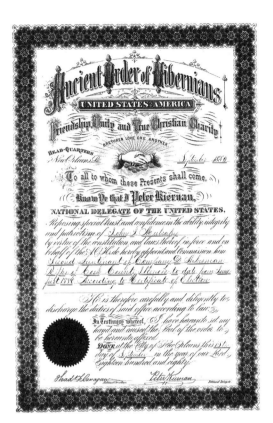

The U.S. branch of the Ancient Order of Hibernians was founded in 1836 in New York. Linked to secret Irish organizations that had fought English tyranny since the 16th century, the AOH in America defended Irish against prejudice and organized St. Patrick's Day parades. This 1880 certificate commissions John J. Mulcahy as a second lieutenant in the Hibernian Rifles, but the military aspect of the AOH was ceremonial by that time.

TIES TO THE OULD SOD

On March 17, 1884, the New York branch of the Friendly Sons of St. Patrick celebrated the organization's centennial. Many New York luminaries attended. The chief speaker was Joseph F. Daly, the president of the Friendly Sons and chief justice of the New York State Supreme Court. His speech described the organization and its functions.

Gentlemen of the Friendly Sons of St. Patrick—I congratulate you on having reached the 100th anniversary of our old society.... We date our society from 1784, but the organization of which it may be said to be a continuance can be traced as far back as 1762, the earliest date that I know of a commemoration of St. Patrick's Day in this city....

When our Society was organized in 1784, among its objects was to find employment for Irish emigrants coming to this city and to relieve them by pecuniary aid in sickness and want. It did this work very effectually until about forty years ago, when the great increase of Irish emigration rendered it impossible to carry out all the purposes for which it was organized, and in consequence, after a great deal of discussion and deliberation, two institutions were formed from the society—the Emigrant Industrial Savings Bank and the Irish Emigrant Society, both of which, upon their separate organization, were composed exclusively of members of the Society—since which period the Society has confined itself solely to discharging, to

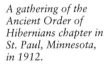

A gathering of the Ancient Order of Hibernians chapter in St. Paul, Minnesota, in 1912.

the extent of its limited ability, the purposes for which it was organized, and celebrated each year by a public banquet its own and the anniversary of the Patron Saint of Ireland.

The standard account of the founding of the Ancient Order of Hibernians is that it was established in New York in 1836 by immigrants who wanted to start a branch in their new country. They wrote to members in Ireland to get a charter. In June 1836 they received the following communication.

From the Brethren in Ireland and Great Britain to the Brethren in New York,

Brothers, Greetings:

Be it known to you and all it may concern that we send to our few brothers in New York full instructions with our authority to establish branches of our Society in America: the qualifications must be as follows:

First: All members must be Roman Catholic and of Irish descent and of good moral character, and none of your members shall join any secret societies contrary to the laws of the Catholic Church and at all times and at all places your motto shall be "Friendship, Unity and True Christian Charity."

You must love without dissimulation, hating evil, cleaving to good, love to one another with brotherly love, without preventing one another, let the love of brotherhood abide in you and forget not hospitality to your immigrant brother that may land on your shores, and we advise you, above all things, have natural Charity among yourselves.

The Irish also organized associations that united people from the same county in Ireland. The first of these was the Sligo Young Men's Association, formed in 1849 New York. Along with social events such as an annual ball, outing, and picnic, membership in an organization enabled members to meet people who might help them achieve greater success and status. Dr. John C. Hannon, a member of the organization, said of the Sligo Association in 1904:

You not only meet friends of your childhood days but you will become associated with men who are higher up in this world than you are and in that way you will have a chance at elevating yourself to some noble point in the history of mankind. When I came to this country, which is not a few years ago, Irishmen at that time had but little education, some none and they were formed for hard work, yes, slaving work, but today you see Irishmen in a different state, they are businessmen, clerks, bookkeepers and in almost every walk of life, hence the necessity of belonging to an Irish County organization whereby you will be enlightened in everything that is going on.

Eamon De Valera

On the night of June 2, 1919, Eamon de Valera slipped aboard a ship in the Liverpool harbor. Having just escaped from a British prison, De Valera headed for the United States. He hoped to raise money from Irish Americans for Ireland's freedom struggle.

Nine days later, De Valera landed in New York, where he had been born in 1882. His Spanish father died when Eamon was an infant, and his Irish mother sent him to her family in County Limerick. There, he became an ardent supporter of Irish independence.

In the short-lived Easter Rebellion of 1916, De Valera led a battalion that captured one of Dublin's government buildings. He was one of the last rebels to surrender. His U.S. citizenship spared him from the death penalty, but he was given a prison sentence.

From his cell, De Valera was elected president of the revolutionary Sinn Fein party. His popularity increased with his dramatic escape and his U.S. tour. Though British police waited to arrest him on his return to Ireland, he slipped through their dragnet and carried on the fight for Irish independence.

De Valera sent representatives to the truce talks with the British in 1921. But he opposed the agreement that resulted, in which the six northern counties of Ulster remained part of Britain. When Ireland's legislature, the Dail Eireann, ratified the treaty, civil war broke out between Irish who supported the treaty and those who opposed it. De Valera was captured and jailed again.

After his release in 1923, De Valera formed a new party, the Fianna Fail. Six years later, he became Ireland's prime minister. Through his efforts, the nation adopted a new constitution in 1937. It declared Ireland a sovereign state called Eire and made Gaelic the official language of the country.

De Valera headed the Irish government for 24 of the 30 years between 1929 and 1959. He kept the nation neutral during World War II, refusing to support the British. In 1959 he was elected to the ceremonial office of president. When he died in 1975, his great dream of a united, independent Ireland remained unfulfilled.

John J. Breen, a member of the Fenian Brotherhood. Established in the late 1850s, the Brotherhood had members in both the United States and Ireland who hoped to organize an Irish rebellion. American Fenians staged two short-lived, unsuccessful invasions of Canada, a British dominion.

It was helpful for a politician to be part of the Irish organizations. James Curley, mayor of Boston, noted:

I had no time for girl friends, since I was kept busy serving as general chairman of committees that organized picnics, outings, minstrel shows, church suppers and dances. As a member of the Ancient Order of Hibernians, I raised funds for welfare projects and went around the neighborhood visiting the sick and needy. In 1896, as in previous years, I served on the general committee that made arrangements for St. Patrick's Day, which culminated in the big parade in South Boston on March 17.

Many Irish Americans regarded their adopted country almost as an extension of Ireland. As Thomas Colley Grattan wrote in 1859:

It is, in fact, unquestionable that the Irishman looks upon America as the refuge of his race, the home of his kindred, the heritage of his children and their children. The Atlantic is, in his mind, less a barrier of separation between land and land than is St. George's Channel. The shores of England are father off, in his heart's geography, than those of New York or Massachusetts.... Ireland—old as she is, and fond as he is of calling her so—seems to him but a part and parcel of that great continent which it sounds, to his notions, unnatural to designate as the *new* world. He has no feeling towards America but that of love and loyalty. To live on her soil, to work for the public good, and die in the country's service, are genuine aspirations of the son of Erin, when he quits the place of his birth for that of his adoption.

Many Irish Americans believed that their low status in the United States was related to the fact that Ireland was not free. Michael Davitt, a leader of a rebellion in Ireland, said in a speech he gave in the United States in 1880:

You want to be honored among the elements that constitute this nation.... You want to be regarded with the respect due you; that you may thus be looked on, aid us in Ireland to remove the stain of degradation from your birth...and (you) will get the respect you deserve.

The Fenians, an organization with members both in Ireland and the United States, worked for Ireland's freedom. Diarmuid O'Donovan Rossa, one of the early Fenians, describes its start in Ireland.

When James Stephens came to Skibbereen in May, 1858, and started the Irish Revolutionary Brotherhood, we commenced to work in that line of labor, and we were not long working, when a great change was noticeable in the temper of the people. In the cellars, in the woods, and on the hillsides, we had our own men drilling in the nighttime, and wars and rumors of wars were on the wings of the wind. The lords and the landlords were becoming visibly alarmed.

Joseph Clarke, another Fenian, described the help the organization received from its branches in the United States.

The idea of establishing an Irish Republic through a world-wide revolt against British rule was a bold conception, which, if carried out at the time and in the manner originally planned had more than a fair chance of success. The main uprising was to take place in Ireland.... All indications were that the revolt would succeed could British reinforcements be held back for any appreciable time. To this end there were to be simultaneous uprisings everywhere throughout the Empire....

Officers from the American army came in twos and threes.... They were forwarded to Ireland in a score of differing ways. As time wore on the risk increased daily. Arrests on suspicion grew frequent.... We used three kinds of invisible ink for mail correspondence. We had a cryptic alphabet for specially important communications. We had drills in the use of weapons.... The dynamite exploits of Captain Mackey, an American officer from the Stephens wing, were conducted entirely outside the local organization.... He disappeared forever in one of his operations on the Thames. [When Clarke himself came under suspicion, he fled to the United States, where he published his memoirs in 1925.]

Though the Fenians failed, Irish Americans continued to support the independence struggle in Ireland until victory was finally achieved in the 20th century. Many regarded it as a sacred duty. Patrick Ford—an ardent nationalist and, as an editor and writer, one of the most important members of the Irish American community—wrote in 1874:

This country is Ireland's base of operations. Here in this Republic—whose flag first flashed on the breeze in defiance of England—whose first national hosts rained an iron hail of destruction upon England's power—here in this land to whose shores English oppression exiled our race—we are free to express the sentiments and to declare the hopes of Ireland. It is your duty, revolutionary chieftains, to realize these hopes! If you are but true to this duty—if you are but true to nature—there are those among you who, perhaps, will yet live to uplift Ireland's banner above the ruins of London, and proclaim with trumpet-tongued voice, whose echoes shall reverberate to the ends of the earth—"The rod of the oppressor is broken! Babylon the great is fallen."

Though Elizabeth Gurley Flynn, the daughter of Irish immigrants, later became one of the leaders of the American Communist party, she never lost sight of her heritage:

The awareness of being Irish came to us as small children, through plaintive song and heroic story. The Irish people fought to wrest their native soil from foreign landlords, to speak their native Gaelic tongue, to worship in the church of their choice, to have their own schools, to be independent and self-governing.... We drew in a burning hatred of British rule with our mother's milk. Until my father died at over eighty, he never said *England* without adding, "God damn her!"

Many Irish American organizations were entirely social in their purposes, such as this Monaghan Men's Irish Dancing Club, photographed in 1905.

Snowy weather does not deter Irish Americans from celebrating their favorite holiday, on March 17.

CHAPTER SIX

PART OF AMERICA

n 1914 President Woodrow Wilson dedicated a monument to John Barry, an Irish immigrant hero of the American Revolution. Wilson pointedly remarked that Barry's "heart crossed the Atlantic with him." Wilson was angry because Irish Americans had protested his administration's policy of selling arms to Great Britain. He said that these new immigrants "need hyphens in their names [as in Irish-American] because only part of them has come over."

Though Wilson himself was the grandson of an Irish Protestant who emigrated from Ulster in 1807, he had little sympathy for the immigrants who arrived later. By questioning their loyalty—and creating the term "hyphenated Americans"—Wilson made it clear that the Catholic Irish Americans had still not been fully accepted.

Yet within 50 years after Wilson's speech, John F. Kennedy stood on the Capitol steps and took the oath of office as the 35th President of the United States. On that day in 1961, seeing one of their own become the first Catholic president, Irish Americans felt they had finally arrived.

It was a triumph that was long in coming. In 1928 the popular Irish American governor of New York, Alfred E. Smith, was defeated in his run for the Presidency—a loss that many felt was due to Smith's Catholic religion. Thirty-two years would pass before the Democratic party, which Irish Americans had supported ever since the days of Thomas Jefferson, nominated another Catholic.

In all other areas of American life, Irish Americans showed how unjust Wilson's belittling judgment had been. In 1917, when Wilson asked Congress to declare war on Germany and its allies (breaking his pledge to keep the country out of war), Irish Americans supported their adopted country loyally and bravely, just as they had in all its previous wars. The 69th Regiment, composed of Irish American troops from New York, was one of the first U.S. units to arrive in Europe. As they had in the Civil War, the members of the Fighting 69th won respect for their valor. The chaplain of the regiment, Father Francis P. Duffy, received medals for heroism from both the United States and France. His statue stands in New York City at Broadway and 43rd Street, in a square named in his honor.

Later, during World War II, the Irish American tradition of military service continued. A Tennessee-born Irish American, Audie Murphy, received more decorations for heroism than any other U.S. soldier. Five brothers from the Sullivan family died together when their ship went down in the Pacific.

Some Irish Americans whose ancestors had arrived on the "coffin ships" of the 1800s reached the height of the American dream. In the year of Wilson's "hyphenated Americans" speech, Henry Ford, the son of a famine immigrant, guaranteed a minimum wage of $5 an hour to the workers in his automobile factories—at a time when the average wage of industrial workers was $11 a *week*. By using assembly-line production, Ford continually lowered the price of his Model T car, making it affordable for millions of Americans. In the process, he transformed American society and made himself one of the richest men in America.

Many other Irish families also acquired great wealth. W. R. Grace, an immigrant from County Cork, was not only New York's first Irish mayor (in 1880) but also the founder of a shipping business that today is a multibillion-dollar company. J. Paul Getty, whose ancestors arrived from County Down in 1824, became a billionaire in the oil business in the 20th century. Joseph P. Kennedy, son of a Boston saloon keeper and father of a President, made an enormous fortune through banking and investments in the 1920s.

The Irish Americans have never lost their talent for politics. More than a dozen descendants of Irish immigrants sit in the U.S. Senate today, and 10 more serve as state governors. Though the old Irish-controlled political machines have declined, Irish Americans still exert considerable influence in big-city governments. Police forces, fire departments, and other city agencies (and their employees' unions) still include many Irish Americans in leadership roles.

Irish Americans, who earlier found the road to success started on the police force, now enter the legal profession. In a field that was formerly dominated by old-line Protestant law firms, Irish American names now appear at the top. In 1981 Sandra Day O'Connor became the first woman to serve on the U.S. Supreme Court.

The Catholic church in the United States, nurtured by Irish priests, nuns, and bishops, now has more members than any other religious denomination. Today there are nearly 60 million American Catholics. And though Irish Americans today make up only about 17 percent of the members of the American Catholic church, nearly half of the bishops still claim an Irish ancestor.

Irish American entertainers have won popularity ever since the early 19th century. Continuing the tradition of the ancient Irish bards, they brought Irish song and dance to their new country. In the 1830s

William Grattan Tyrone Power toured the country with his stage troupe. More than a century later, his great-grandson Tyrone Power became a leading man in Hollywood movies. Another great-grandson, Tyrone Guthrie, founded and directed a distinguished theater company in Minneapolis.

The operettas of Victor Herbert,

Irish Americans in Boise, Idaho, were represented by this float in the town's Wild West Stampede in 1913. The light bulbs outlining the large shamrock were a modern touch, though the float was drawn by a team of horses.

who arrived from Dublin in 1886, have charmed Americans ever since. In that same era vaudeville duos such as Gallagher and Sheehan and Harrigan and Hart regaled audiences with their slapstick stereotypes of Irish American immigrants.

After the turn of the century, George Michael Cohan (whose family name was originally Keohane) created a new stage image for the Irish American. Cohan's patriotic songs, written for the Broadway plays in which he starred, became perennial favorites. "Over There," a song he created to

send U.S. troops off to Europe in World War I, earned Cohan a special medal from the U.S. Congress. As portrayed by another Irish American actor, James Cagney, in the 1943 movie *Yankee Doodle Dandy,* Cohan lives on in television reruns.

Countless other Irish Americans attained stardom in the movies, helping to break down anti-Catholic prejudice. In the 1940s Bing Crosby and Barry Fitzgerald starred as lovable priests in *Going My Way* and *The Bells of St. Mary's.* Pat O'Brien and Spencer Tracy played many roles as priests, war heroes, or football coaches. Gene Kelly and Donald O'Connor continued the Irish dance tradition in joyous Hollywood musicals.

Director John Ford explored the tragedy of Irish rebellion in his Academy Award–winning film *The Informer.* Ford also depicted Irish village life in *The Quiet Man,* starring John Wayne and Maureen O'Hara. Wayne (born Marion Michael Morrison) became a bigger-than-life screen hero in Hollywood's version of the Old West. Grace Kelly, daughter of a wealthy Philadelphia family, enchanted movie audiences with her classic Irish beauty before becoming a real-life princess when she married Prince Rainier of Monaco.

Today's Irish American actors—such as Michael Keaton, Jack Nicholson, and Sean Penn—no longer need to play "Irish" roles. Audiences all over the world view them simply as Americans—

another sign of the assimilation of Irish Americans into the mainstream.

After television arrived in the 1950s, Irish American stars appeared in virtually every American home. Among the most popular were Jackie Gleason and Art Carney, who played a hot-tempered bus driver and a sewer worker in *The Honeymooners*. Ralph Kramden and Ed Norton, the characters they played, reflected the experiences of many Irish Americans, including the actors themselves. In the 1970s Carroll O'Connor and Maureen Stapleton created another popular TV version of lower-middle-class urban American life in *All in the Family*.

Irish American playwrights and authors carried on the long literary tradition of Ireland. Eugene O'Neill, the only American playwright to win the Nobel Prize, was the son of an Irish immigrant.

James T. Farrell, in his 1932 novel *Studs Lonigan,* depicted an Irish American youth in a poor neighborhood of Chicago. Farrell's stark, gritty style looked unblinkingly at the effects of poverty. However, his works never became as popular as those of three other Irish American novelists, F. Scott Fitzgerald, Margaret Mitchell, and John O'Hara. The narrator of O'Hara's first novel says: "I want to tell you something about myself that will help to explain a lot of things about me...."

First of all I am a Mick." Fitzgerald's best-known novel, *The Great Gatsby*, is the story of a man born into poverty who gains a fortune but fails to win the love of an upper-class woman. Though Gatsby is not Irish, his struggle could be that of many Irish Americans who found that money was

Many Irish Americans hear the phrase, "The map of Ireland is written on your face." That distinctive Irish smile comes from the heritage of Gaels, Celts, and Normans in the land known as the "emerald isle."

not necessarily the ticket to upper-class society.

In her Civil War epic, *Gone with the Wind*, Mitchell created one of the most popular heroines in American fiction—Scarlett O'Hara, whose family owns the Southern plantation called Tara. Mitchell was drawing on her own Irish roots. Her grandfather, Thomas Fitzgerald, was a plantation owner in Georgia before the war.

Later Irish American writers have looked to their roots for inspiration. Edwin O'Connor, in *The Last Hurrah,* gave a balanced portrayal of a Boston Irish mayor—

clearly modeled after James Michael Curley—who is both rogue and hero. Mary Deasy, in numerous short stories, drew on her experiences growing up in a multitalented Irish American family. Today, William Kennedy is continuing his award-winning series of novels about Irish Americans in Albany, New York.

Irish Americans have kept their affection for the land their ancestors called home. Irish music, sung by the Clancy Brothers, Tommy Makem, and other artists, remains popular. There is even an Irish American rock group called Black 47 (after the worst year of the famine) that reaches young audiences on MTV. Gaelic societies preserve the knowledge of ancient Irish language and tradition.

No Irish American festival, of course, is as popular as St. Patrick's Day. The first St. Patrick's Day parade in New York City took place in 1759. More than two centuries later, the green banners of Irish counties, clubs, schools, and churches are still unfurled each March 17 for the march down Fifth Avenue. The drums roll and bagpipes skirl as the marchers move forward.

All over the United States, in cities large and small, similar celebrations take place as the Irish assert their affection for the land their ancestors called home. Countless other Americans, with not a drop of Irish blood in their veins, join them. For it is a day when everyone celebrates the triumphant history of the Irish in America.

The inauguration of John Fitzgerald Kennedy as President of the United States in January 1961 was the moment when Catholic Irish Americans felt that at last they had won full equality in the United States.

MAKING IT

John F. Kennedy squarely faced the "Catholic issue" during his campaign for the Presidency in 1960. In a speech before Protestant ministers in Houston, Texas, he declared:

Because I am a Catholic, and no Catholic has ever been elected President, the real issues in this campaign have been obscured.... So it is apparently necessary for me to state once again, not what kind of church I believe in, for that should be important only to me, but what kind of America I believe in.

I believe in an America where the separation of church and state is absolute....

I believe in an America where religious intolerance will someday end....

I believe in a President whose views on religion are his own private affair....

This is the kind of America I believe in—and this is the kind of America I fought for in the South Pacific and the kind my brother died for in Europe. No one suggested then that we might have a "divided loyalty," that we did "not believe in liberty" or that we belonged to a disloyal group that threatened "the freedoms for which our forefathers died."...

Contrary to common newspaper usage, I am not the Catholic candidate for President. I am the Democratic Party's candidate for President who happens also to be Catholic. I do not speak for my church on public matters—and the church does not speak for me....

If this election is decided on the basis that 40 million Americans lost their chance of being President on the day they were baptized, then it is the whole nation that will be the loser in the eyes of Catholics and non-Catholics around the world, in the eyes of history, and in the eyes of our own people.

Kennedy won the election. His victory indicated to many Irish Americans that they were at last fully accepted. Thus, when Kennedy was assassinated, they felt a tremendous personal loss. The Irish American sportswriter Jimmy Cannon summed up the feeling.

It would be impertinent for me to measure John Fitzgerald Kennedy on the sports pages. The historians will do that, and in other parts of the paper qualified reporters will describe him as a public man....

But I am my father's son, and John Fitzgerald Kennedy's death seems a private matter of my family. Never did I meet him, but I felt close to him—he was never a stranger....

John Fitzgerald Kennedy would have been no alien on the tenement-sad streets of my childhood where my people still live and mourn today. He was rich by inheritance, an educated man, but his ancestors suffered as we did. He was the perfection of our breed, and my affection for him was not influenced by political reasons, although my old man was a Tammany captain in a ward where Republicans were regarded as rebellious cranks. I regret my father died before John Fitzgerald Kennedy was elected president because he had his heart broken when Alfred E. Smith was beaten.

My old man felt betrayed by his country and went to his grave believing no man who was an Irish Catholic could ever be president. It is indecently improper to vote for a man because of his racial origin or his religion as it is to turn against him for the same motives. But my old man brought me up to be proud of what I am. He never forgave anyone who objected to Al Smith for being an Irish Catholic and he denounced them as vandals of patriotism. It would have reinstated my old man's faith in his countrymen to know they accepted John Fitzgerald Kennedy on his reputation as a congressman and a senator....

He was the best of us and when he made it to the White House, we were no longer Micks. Of course, we had other big men, but what happened to Al Smith made their accomplishments diminish until John Fitzgerald Kennedy proved that we were true citizens of our country, and not just tolerated intruders. It may sound exaggerated, but I lived with this....

Other reporters will describe how tall he stood against the horizon of history. I only know that Boston voice always seemed to be talking to me and mine. Every time I heard it, I felt there was a message in it for my dead father. And for all those who took Al Smith's disappointment as a family tragedy. They weep for John Fitzgerald Kennedy today, and for him their candles burn.

Mary Higgins Clark is the author of numerous best-selling novels. The success she enjoys today came after many years of struggle, however. Clark recalled how her mother, Nora, the daughter of Irish immigrants, raised three children in the face of hardship and tragedy.

All her life, Mother had dreamt of owning a home of her own and she and my father bought one a few years after they were married [in the late 1920s]. To Mother, Buckingham Palace, the Taj Mahal and Shangri-La were all wrapped up in that six room, brick, semi-detached dwelling in the Pelham Parkway section of the Bronx. But then the depression years set in. My father's once-flourishing Irish Pub began to lose money. Their stocks were lost in the crash; their savings dwindled to nothing.

[After Clark's father died in 1939, her mother took in boarders to earn money.] In spite of all our concerted efforts, the roomers who came and went, our babysitting jobs and Joe's [Clark's older brother] newspaper route, Mother couldn't

Sandra Day O'Connor

On September 26, 1981, Sandra Day O'Connor took the oath of office as a justice of the Supreme Court of the United States. It was a historic moment, for O'Connor became the first woman ever to serve on the Court.

She is the descendant of immigrants who left Ireland during the famine. Her great-grandparents spelled their name O'Dea, but their son changed it to Day. Around 1880 Henry Clay Day left Vermont, where his parents had settled, and went west. He started a cattle ranch in the Arizona Territory. At his death in 1921, his youngest son Harry took over the ranch. Harry married Ada Mae Wilkey in 1928, and Sandra was born two years later.

The ranch was in an isolated area with no nearby schools, and Sandra's mother taught her to read. When she was five, Sandra was sent to Texas to attend school. Homesick, she returned to the ranch at 13 and made a 22-mile trip twice a day to high school. Graduating at 16, she enrolled in Stanford University and received her law degree there in 1952.

Soon after, she married a classmate, John Jay O'Connor III. Sandra applied to law firms for a job, but, as she recalled, "None had ever hired a woman before," and they turned her down. Thus, she entered public service as a county attorney in San Mateo, California.

The O'Connors moved to Arizona in 1957. Sandra opened her own law firm so that she could allow time for raising her three sons. In 1969 she became a Republican member of the Arizona senate. Three years later her colleagues elected her majority leader—the first woman to hold that office in any state senate in the nation.

In 1974 O'Connor won election as a county judge. Five years later Arizona's governor appointed her to the state court of appeals. Her even-handed decisions won the support of liberals and conservatives when President Ronald Reagan named her to the nation's highest court. In her confirmation hearings, when asked how she wanted to be remembered, she replied, "Ah, the tombstone question. I hope it says, 'Here lies a good judge.'"

Members of the Brennan family of Oak Park, Illinois, display their Easter outfits in 1961. Thomas Brennan, the father, personally made the dresses and suits for his six daughters and six sons.

keep up the mortgage payments and lost the house. She was urged to take Joseph out of school and put him to work but she refused. "Education is more important than any house," she said firmly. "Joseph will get his diploma." Our next stop was a three-room apartment near the trolley line and into it she moved the full contents of the six rooms, sure that someday our fortunes would change and we'd get the house back. We never did, and whenever she returned from visiting the old neighborhood, her eyes would shine with unshed tears as she remarked how beautifully her roses had grown....

At thirteen, Joe contracted osteomyelitis [a bone disease]. Mother was told that an operation to remove the hipbone was necessary to save his life. Widowed only a few months she made the stunning decision not to operate. She wouldn't make a cripple of Joseph and she knew God wouldn't take him from her. It was Christmas.... Mother and John [Clark's younger brother] and I carried all his presents to the hospital. His main gift was a hockey stick. "You'll use it next year," she promised him. He did.

Joe graduated from high school in 1944. Mother could have claimed him as her sole support and kept him out of service. Instead she let him enlist in the navy with his friends. Six months later she took the only long trip of her life, a plane ride to California to be at Joe's deathbed in the Long Beach Naval Hospital. To the people who fumbled for words of sympathy she said, "It is God's will. I couldn't let Joseph go when he was sick the other time but now God wants him even more than I do."...

Her pride in all of us was enormous.... When John went to Notre Dame she must have written a dozen letters to long-forgotten cousins. The letters began, "My, what a busy summer, what with getting John ready for Notre Dame...."

[When Clark's mother died,] she had a total of $1,700 in insurance from nickel and dime policies she'd paid on for years. They were tied together in an old brown envelope. There was a note to Johnny and me with them. It said, "Don't waste more than a thousand dollars on the funeral. Give one hundred dollars to each of my grandchildren." She didn't realize that she'd already given us all a priceless legacy, her ceaseless devotion and unfailing love....

[Nora Higgins did not live to see her daughter become a successful novelist, but Mary Higgins Clark believes that her mother had something to with it.] I have mother's old black felt hat with the brief edging of black veiling in my closet. It's battered now and out of shape but over the years when things weren't going well, when the bills were piling up or one of the children was sick, I'd give it a quick rub and say, "Come on, Nora, do your stuff." I had no doubt that my first novel would be successful because it was dedicated to her. "I can just see Nora," a friend said laughing, "Dear Lord, not to bother you...the paperback sale on the book was excellent but *how about the movie rights?*"

Mary Kenny arrived in New York in the 1960s and got married soon afterward. She described how she and her husband attained success in the United States.

After we married we settled in an apartment on Fordham Road in the Bronx. We stayed there for two years; then we moved down to 161st Street, right beside Yankee Stadium.... I had my first three children—Patricia, Caroline and Siobhan—close together. They were born within the first four years of our marriage and it was tough trying to raise them. Patrick worked and I had to take care of them. We didn't get out much because we didn't have relatives to babysit. We couldn't pay a babysitter. When we came back from our honeymoon, I think we had a hundred dollars or something in the bank after paying for the wedding and the furniture. But we managed somehow. God provided....

After four years there, we decided to invest in a house. My husband worked at a brokerage house down in the Wall Street area. We had saved a few dollars and invested in some stocks and made what you might call a killing on the market. We finally had enough money to put a down payment on a house....

We have been fortunate in America. Five years ago my husband and I both became American citizens. We decided that America had been very good to us. Thank God! We've made a good living. My husband has never been out of work. I feel that it's been important for me to stay home and raise the children.

Still, I think there might have been some prejudice, some coolness towards us because we were Irish. But I think at some time or another all groups have had some prejudice directed at them. I learned that at one time the Irish were told they need not apply for certain jobs. That's different today. I think people take you more or less for what you are. I don't think they label you as Irish, German or Italian as if it's something bad.

For after all, this is a land of very few native Americans. We all immigrated from other countries. I think it makes life so much more interesting to have all these different nationalities. My cousin Anne is married to an Italian. We love him a lot. He's come down to our home and eaten cabbage and turnips for the first time in his life. We go to visit them and he's introduced us to antipasto and lasagna. We get a great kick out of this. Where else on earth could you find this? Only in America could you find so many different nationalities.

This is the land of opportunity. I have found that out. Pretty much whatever you want to be here you can become. There are a lot more opportunities here than in Ireland. You can always go back to school, as I have, to finish your education. The opportunities are here.

So, over all, America has been great to us and we owe it a lot. If I had the chance to come here again, to do it all over again, I would.

Georgia O'Keeffe

A 1922 portrait of O'Keeffe by Alfred Stieglitz.

Most Americans know Georgia O'Keeffe's work from her stark paintings of cattle skulls and vividly colored desert flowers. But her long career encompassed many of the styles of 20th-century art. Her first show took place in New York in 1916, and she continued working till her death in Santa Fe in 1986. During those decades, she became legendary in the world of modern art.

O'Keeffe was born on a wheat farm in Wisconsin in 1887. Her father was an Irish immigrant who encouraged his daughter's talent. Georgia attended a nearby convent school, and when she was 17 enrolled in the Art Institute of Chicago. After graduation, she supported herself by teaching. Restless, she moved from Chicago to South Carolina and then to Texas.

O'Keeffe began to make abstract drawings that drew on her memories of the prairies of her native Midwest. Alfred Stieglitz, a photographer who had opened an art gallery in New York dedicated to the work of young Americans, arranged for her first exhibition. The resulting acclaim enabled her to take up painting full-time. Awed by New York's skyscrapers, she painted enormous canvases, saying "I'll make them big like the huge buildings.... People will be startled; they'll have to look at them."

In 1924 O'Keeffe and Stieglitz married. He was fascinated by her beauty, and as she recalled late in her life, "He photographed me until I was crazy." After Stieglitz died in 1946, O'Keeffe moved to New Mexico. For nearly a quarter of a century, she worked in near obscurity. Then in 1970, when she was 83, a New York museum held an exhibition of her new paintings. The art world was stunned, for they were nothing like the then fashionable work of other artists. O'Keeffe once again became a celebrity, but she shunned the city for the New Mexico deserts, which had inspired her greatest work.

NEW IMMIGRANTS

The United States remains a place of refuge for Irish today. Many new immigrants are attracted by the success stories of friends or relatives who are already here. Ann Gilvarry, who came to New York when she was 17, explained:

I don't remember when I first started thinking about emigrating. Probably I was very young. My uncle had come over here. He and my aunt lived in the Bronx and when they came back to visit us they seemed to have a good life. They would rent a car—something that seemed very extravagant to us. They just seemed to have plenty of money for whatever they wanted to do. Not rich. I didn't think they were rich but I knew that they were very comfortable.

I would have gone someplace else if not to America. Maybe Dublin or Galway. I knew I never wanted to stay on the farm. My father had seventy-five acres. Started out with only twelve but by the time he died he had seventy-five. The neighbors used to say, "He had a nice bit of land." But everybody knew my brother would get that. Nobody ever thought different or even questioned it. So the four of us (I have three sisters) we all came to America.

What did my parents say? Well, they couldn't object. Mind you I told them I was leaving but I didn't ask their permission.

Today, Irish immigrants start their voyage to America by taking a ferry to an airport, where they will board a transatlantic jet. A journey that once took weeks is now only a few hours long—and much more comfortable.

My aunt sponsored me so my parents knew I was going to be looked after. One sister was already here so it seemed more like I was going to join family.

Some immigrants find the cultural differences between Ireland and the United States shocking. Maura Mulligan came to the United States in the 1970s. Nearly 20 years later, she described her experiences to an interviewer.

When I first arrived here at sixteen, everything seemed too big—even the people. It was somewhat overwhelming. My first job here was as a long-distance operator. I was living in Queens [New York] with my uncle and commuting to Manhattan. I was instructed not to look at anyone on the street or speak to strangers. After living in a small village where everyone knew me, it was a hard adjustment, but I liked my job with the phone company. My home away from home was the Irish dancing school that I joined. It helped me maintain my identity in the world I left behind. Even after five years, I couldn't adjust to American culture—especially the men. I couldn't stand the way they looked at us, as if we were meat, when I would go to dances. I didn't think I'd ever adjust, so I retreated by joining a convent. I wore a habit for seventeen years. In the convent I got my high school and college degrees and got my teaching certificate. Over the years I came to realize that I couldn't continue my personal growth there—that I had to wake up and rejoin the world. Leaving the convent was almost like getting a divorce. It wasn't an impulse move—I thought about it for five years before I actually left.

The world has changed a lot in the last seventeen years. I went into therapy to begin to catch up and figure things out. The most interesting thing I've discovered is that I'm not afraid of people anymore. I'm no longer that Irish farm girl who was afraid to make eye contact on the subway. I'm able to go out on dates and enjoy men's company. I'm more comfortable with myself.... I've opened my own Irish language school, and I find great personal satisfaction running my own business. I'm very excited about my future. I'm enjoying the music and dance of Ireland again, but I wouldn't want to go back to Ireland to live because I've gotten used to an American open-mindedness that only can exist in a country this large.... If I joined a convent in Ireland, I probably never would have left, but here the atmosphere is much more conducive to thought and change. I certainly would never have gone into therapy—that's a very American thing to do, to question and ask why. This is a wonderful country for the curious. I'm not just running on faith anymore. It's time for me to ask questions and find out who I am. I'm glad I had the courage to leave Ireland and then the convent. It makes me who I am today, and I'm enjoying getting to know myself better and I like who I am.

These five sisters of one family, all of whom became nuns, were born in County Galway and immigrated to the United States in the 1920s.

Mary Branley, who emigrated from Ireland in the 1980s, now teaches at the Cathedral Grammar School in Boston.

Pat Tierney is a roaming street poet and ardent Irish nationalist. Yet he, too, came to the United States in search of work.

When I was seventeen I went to Dublin. Places like the Garden of Remembrance, the G.P.O., Kilmainham Gaol [reminders of the Easter Rebellion of 1916] had a strong effect on me. I liked Dublin. I knocked around and wrote various little poems. When I was eighteen I joined Sinn Fein and became a recruiting officer. Then I was arrested for various political activities. Like I burned a Union Jack in front of the St. Patrick's Day reviewing stand. Then we got into selling books on the street at the G.P.O. and I remember this one man mentioning Yeats [William Butler Yeats, generally considered the greatest Irish poet of the 20th century] and sometimes he'd quote a line here or there. The people I met in the Republican circles were the ones who really made me literate and gave me the sense that I could achieve things myself. Then I decided I wanted to go to America. I saved money and went over on holiday but, of course, remained illegally. I went to Detroit and worked as a janitor, then went to Arizona and met some Irish people there and started to organise poetry readings. Then I went to Massachusetts to work on fishing boats and went to Wyoming and worked in an oil field and on a ranch during branding season. I was accumulating a larger wealth of experience to draw from...I knew I wanted to get to poetry.

Ann Gilvarry came to New York in 1968. She describes how she gradually decided to stay for good.

My first job in New York was taking care of kids. Good job. Roof over your head and fifty dollars a week. All I thought about was saving up for the trip back to Ireland in the summer. And that's what I did: worked until June, quit my job and went back to Ireland for three months. Then back to New York to find a new family to work for. Same thing the next year. After about five years of that I decided I'd do something else. Get a different kind of job for one thing. The last woman whose children I watched was a little crazy and that's what made me decide to go to beautician's school. Now I work on 53rd Street here in Manhattan. Make a nice little bit of money. Pretty much can do as I please. Quite a few Irish girls go to beauty school—not the illegal ones, they have to work for families off the books—but those of us who have green cards, yes.

I'm not quite sure exactly why I chose this kind of work or even when it was I decided I would probably stay in America permanently. Those summers back in Ireland just stopped. I guess in 1975. One thing—I had become engaged to a boy there but then we broke off. So really the reason for going back just wasn't there. And I became an American citizen. My younger sister had come to New York and my life seemed more here than there.

Like so many who came earlier, today's immigrants often retain ties to their native land. Teresa O'Hara was a special case. She was born in New York in the 1950s to immigrant parents. When Teresa was very young, her family returned to County Galway. There she received a good education, earning a degree in social work and then attending the London School of Economics.

Finding no market for her skills and education in Ireland, Teresa O'Hara returned to the United States in 1987. Because she is a native-born citizen, she could enter the country freely. Since then, she has become active in the Irish Immigration Reform Movement (IIRM), which was founded by Irish immigrants.

I did feel I was privileged in a sense that with my...American citizenship, I could come here and choose what work I wanted to do. I really felt for people that came here...and would like to get work experience, but were unable to do so because they hadn't the proper documentation. I did get involved in different committees of the Irish Immigration Reform Movement, the Political Action Committee. I didn't become involved in the main committee until this year, January of 1989. I know the goals [of the organization] are very good and legitimate in terms of trying to work on legislation [to help immigrants], but for me in some ways it was a bit vague and not something I could put a lot into. At the end of last year they started developing the [telephone] hot line and trying to have more outreach to immigrants. I got involved developing the social aspects of the group.

[O'Hara felt that the male members tended to dominate the group.] I can make decisions as well as the men on the group and I'm as well able to fight and stand up for the rights of young Irish immigrants. It's just that I would go about it in a different way....

In terms of the IIRM, I try to get different people in actually running the events. We're going to develop a whole lot more on the social side of things, networking with other groups. We have a radio show going now and I'd like to see more people getting something out of it.

The majority talk about returning to Ireland eventually. It's something amazing, very, very few talk about staying here long term. I do know a few people who do, people who have met people over here and they're going to get married. Most people have this vision of eventually settling back home, saving up the money for the time when they return to Ireland. I think things are changing a bit at home, but it does take having dynamic people who are going to get involved in changing the system. It won't be easy at first. You have to have a vision for the country.... I have this vision of Ireland at the moment that it doesn't have a young population, that they're all leaving. That's why I see the need for emigrants coming back. These are the people that can help change the country.

At Cleveland's Irish American Cultural Festival, musicians play ballads and folk songs, both modern and traditional.

From the beginning, the celebrations of Saint Patrick's Day were associated with patriotism for the Irish Americans' adopted country as well. On March 18, 1863, during the Civil War, the New York Herald *commented on the St. Patrick's Day festivities held the day before:*

The celebration yesterday was, in every point of view, a Union demonstration. The great lesson of the day to every Irishman was: Stand by the country which gives you life, liberty and right to be happy in your own way. We do not doubt that this lesson was learned, and will be remembered by many a patriotic Celt. The influence of St. Patrick's day, and especially of St. Patrick's night, will last, not only for to-day, but, we hope, for many days to come. We congratulate our Irish fellow citizens upon the fine weather, the large turnout and the splendid procession, and particularly upon the good order and decorum which marked all the proceedings.

The traditional Irish dance called step dancing is displayed by this group of girls in a St. Patrick's Day parade in Chicago in 1992.

ERIN GO BRAGH

A favorite Irish greeting is Erin go bragh, *which means "Ireland forever." Irish Americans have proudly preserved their sense of Irishness, along with many of the traditions of their homeland. Gaelic-language classes, Irish dance and theater groups, and Irish studies departments in colleges are all evidence that Irish Americans' cultural ties to their homeland are still strong.*

The greatest Irish American feast of all is St. Patrick's Day, March 17, when celebrations are held throughout the United States. William O'Connell, later to head the archdiocese of Boston as a cardinal, grew up in Lowell, Massachusetts, in the 1860s and 1870s. In his autobiography, Cardinal O'Connell recalled the significance of the St. Patrick's Day parade in Lowell.

One of the things...which helped to demonstrate the growing strength of the Catholic population in the city was the parade of the Irish Catholic men on the great feast of their patron saint. Early in the morning of St. Patrick's Day the people flocked to their churches.... After the High Mass the great procession was formed and, headed by bands and the floating banners of America and Ireland, thousands of men paraded through the principal streets, while the rest of the population looked on with mingled sentiments, the uppermost of which was surprise at the number and appearance of those men and boys, their faces aglow as they marched with sturdy stride over the long route. Here and there one caught a smile of derision on the faces of some of the onlookers. No doubt, to many the sight was not a welcome one.... But the marchers, young and old, with heads erect and eyes looking straight ahead, as if into the future, felt that already the time had at last arrived when their existence, their power, and their numbers must be reckoned with. The bands played all along the route the wonderful marching tunes of the old Irish melodies, the most spirited of which was "Garry Owen".... Evening came and the celebration was continued at great public banquets, where in vivid oratory, natural to the Irish, were recalled all the glories, the triumphs, and the sufferings of the Irish race.

To the chagrin of the mill-owners and managers, the workers made the seventeenth of March a great holiday, and to their wonder, and one might say their anxiety as well, the machinery had to be stopped and the mill gates closed. When one remembers that this was not done in those times even on Christmas Day...the full significance of the event, I say, will be quickly realized. The Irish Catholics of Lowell, by dint of vigorous, passive resistance, were beginning to win their rightful place and to force a reluctant recognition from those who hitherto had utterly ignored them and their human rights as citizens and workers.

Catherine Hoy remembered St. Patrick's Day in Butte, Montana, in the early years of the 20th century.

St. Patrick's Day. That was the day of all days. We would all go to church, to Mass. We'd have these great big green bows in our hair, you know. We'd be all decked out in big shamrocks and so on. Then we'd all march. Oh, sometimes it was nice weather, and sometimes you were knee-deep in slush and mud. You had to wear overshoes, but still you had the big green bow in your hair. We'd march all around town. My father was a Hibernian and he wore this big regalia, you know. Then we'd go back to the church, the basement of the church, and we'd have a banquet, food, as I said, that you wouldn't believe. Everybody brought something. Two or three hundred people. There was food until you couldn't...well, just food, food, food. And all the beer you wanted. Kegs of beer. There were a lot of them rolled home, but a lot of them made it all right.

Nellie T. O'Donoghue O'Leary, who arrived in the United States in 1920 from the village of Rathmore in County Kerry, kept up the Irish tradition of sharing with neighbors. As her daughter, Judy O'Leary Anderson, recalled:

My mother was a remarkable woman who was loved and admired by everyone who knew her. Hers was the typical Irish story: a marriage arranged for the only son in the family, necessitating in the three daughters leaving the farm to find positions elsewhere. She was sponsored by an uncle and immigrated to the U.S.A....

From an early age on, we knew that our mother was someone special. In the early 1930s, it wasn't unusual to walk into our back hall and see a homeless victim of the Depression eating on the steps.

Every holiday we would have one or two children from a local orphanage joining in our celebration.... More often than not we'd have an elderly neighbor too, and many's the time we'd have to deliver a basket with a hot turkey dinner in it to an ill or confined neighbor before we sat down to our own. We'd take turns doing it. There were six children in our family, but there was always enough food to share.

Mom was always there when there was a need. During the war she rolled bandages and worked for the Ladies of Charity, sewing and making clothes for poor children. She was the one who walked door-to-door, collecting money for the March of Dimes and signing up neighbors to give blood and/or donations for other worthwhile organizations. If there was a fund-raiser for the church or school she was either in the kitchen cooking the dinner or working at the bake sale or card party....

Every St. Patrick's Day, after the parade, we would have about thirty people squeezed into our little flat for a ham and corned beef and cabbage dinner with Irish Sodabread and tea.

NELLIE O'LEARY'S IRISH SODABREAD

Judy O'Leary Anderson treasured a recipe for Irish sodabread made by her mother, Nellie T. O'Donoghue O'Leary, who arrived at Ellis Island in 1920 from County Kerry:

INGREDIENTS

4 cups all-purpose flour
1 cup sugar
4 tsp. baking powder
1/2 tsp. salt
1 stick melted butter
1 1/2 cups raisins
2 tbsp. caraway seeds
1 1/2 cups buttermilk
1 egg, slightly beaten
1/3 tsp. baking soda

Sift flour, salt, baking powder, and sugar; add melted butter and mix. Stir in raisins and caraway seeds. Combine buttermilk, egg, and baking soda. Make a well in the center of the batter. Pour liquid ingredients and stir into flour mixture. Place in large iron frying pan, well-buttered. Use a knife to make a cross on the top. Moisten with melted butter. Bake in a 375 degree oven for an hour, or until golden brown and shrinks from the side of the pan.

Because so many Irish Americans still have relatives in Ireland, ties between the two communities have remained strong. Kitty Fitzgerald Donovan was born in County Limerick and came to the United States as a 19-year-old in 1927. Today she lives in Oak Park, a suburb of Chicago, but has often returned to Ireland.

For the last 20 years every summer I've gone back. I took my oldest grandchild, Sarah, when she was 10 and a half. For six years after that she went and just loved it. I suppose she liked it because it [Newcastle, County Limerick] was so much like a small town....

My husband was born in this country. But his folks were from County Cork. He always wanted to go back and see where his folks had lived. We took his sister Jo with us one year. She knew so much about the old home from the stories her mother told her. She identified a panel on the wall. The people who lived there now were her people. Many of the people had faces so much like my husband's. Through the trips, I keep in touch with my family and Ireland. Last year, I went to Ireland and then to London to see my sister who is an Ursuline nun. She was told she had cancer and only had one month to live. I went right away. It was sad saying goodbye. Unfortunately the doctor was right.

Fionnuala McKenna is the founder and head of the New Irish Theatre Company in Boston. Although McKenna arrived in the United States in 1987 without a green card (a work permit), she had boundless energy and optimism. She described how she came to establish an Irish theater.

I arrived in America in January of 1987.... God Almighty, I was knee deep in snow.... It wasn't a great apartment either. I just bunked on the floor. I think I had $400 and I automatically assumed that within a week I was going to get a job. Little did I know that three months later I still wouldn't have a job. I thought I was depressed before, but I never was as near to having a nervous breakdown.... For the first time in my life I really was disappointed with myself and my own judgement....

There was just a void here.... The whole social thing was very different and I was missing something. I thought, "Wouldn't it be great to put on a play?"... In the matter of a few weeks I got about fifteen names together. I got someone who was interested in directing. Nobody had any experience at all. They were all Irish immigrants, mostly illegal. So, through trial and error we managed to pull ourselves together. By the following April we had the *Juno and the Paycock* [by the Irish playwright Sean O'Casey] in production. It was great—everybody who came to see it was blown away by it....

I must say America has released a valve in me, just like the steam of a kettle.... Whether it's just the isolation of being totally on my own, independent, away from family has done it or whether it's the struggle involved with being here and having to fight for survival, has caused me to be more determined.... I

Bagpipers, playing the ancient songs that once sent Irish soldiers into battle, are now a regular part of Irish American celebrations.

suppose there are many factors really. I don't sense the depression here that I do at home. I felt strangled at home. I was just held back.

Kevin M. Cahill, a physician, is the director general of the American Irish Historical Society. In a speech at the society's annual banquet, he recalled the importance of Irish heroes—both ancient and modern—in his childhood.

Erin go Bragh *is Gaelic for "Ireland Forever." The phrase is repeated wherever Irish Americans gather to celebrate their heritage.*

When I was a young boy, my father filled the house with tales of Irish heroes. Since the Cahills are direct descendants of the High Kings of Ireland, there was, naturally enough, a modest emphasis on our own genetic strain. In all fairness, I should note that the land of our forebears was blessed with an inordinate number of High Kings; the rest of the population were saints, scholars, poets, and patriots, although there was constant argument as to who belonged in the latter group.

After a series of reverses, the Court of the Cahill royal line moved from some misty mound in Ireland to a picturesque area just off Fordham Road in the Bronx. There, in cold water apartments, affectionately known as the Kerry flats, were the heroes we were taught to admire. There was nothing mythical about them—they were Irish immigrants.

I could never quite distinguish between the blood relatives and the boat relatives, those who had shared the common bond of surviving together the trail to Ellis Island. To a youngster, these heroes were sometimes overwhelming—they certainly, on occasion, drank too much and fought—mostly with words— and struggled, usually unsuccessfully, to shed the past too quickly. But far more impressive was how deeply they loved the clan, and they were loyal almost to a fault. There was nothing subtle about their dreams—they had come to this land so that their children could enjoy a better life. These American-Irish heroes did not leave the soft, gentle Ireland celebrated in song. The wealthy landowner and the city gentleman did not leave—our ancestors were the survivors of oppression and famine, those desperate enough to seek refuge in steerage. But when their spirit burst forth upon this great land of the United States, they and the nation prospered.

Two Irish bagpipers in Dawson City, Alaska. During the Alaska gold rush of 1898, many Irish Americans headed to the Far North to seek their fortune. In 1908, Martin Gateley, an Irish miner who did strike it rich in Alaska, sent a letter to the British government offering to buy Ireland from England. He was turned down.

ITA DONAHUE

Ita Donahue's house in Salem, New Hampshire, is unmistakably the home of an Irish American family. Banners with green shamrocks and Gaelic words of welcome hang from the roof. Of course, this is a special weekend—St. Patrick's Day. On Sunday Irish Americans from all over the area will march in the annual parade through nearby Lawrence, Massachusetts, and Donahue's household is thoroughly involved in the arrangements.

Ita Donahue:

I've put some tea on, and while it's brewing, we can look at some of the pictures of my family and talk about the memories they contain. This teapot is the Irish china—Belleek. They have a certain type of potash that is found only near the town of Belleek, near Donegal.

Here are some real Irish shamrocks in a pot, that I've brought back and kept for years. Now these are large, and we have a great debate over what is a shamrock and what isn't. Those that come out of the ground in March, on St. Patrick's Day, are quite small, and some claim those are the only true shamrocks. But these have grown over the years since I brought them from Ireland.

This is a picture of my dad, Lawrence Justice Sullivan. He came from Newmarket, a typical Irish town in north County Cork. When I went back with my mother for a visit in 1970, I had to find the town myself, for Dad was gone by then. But I remembered so well him telling me as a little girl, "The road is lined with flowers as you ride into town." And I found the road. Roses and flowers are everywhere. Rhododendrons cover the hillside and vales overhanging the lakes of Killarney in May and June.

Enough on the flowers in dad's memory. He was born in 1877, the youngest of six children. His dad, Richard Sullivan, was an officer in the English army way back then, and his mom was Jane Justice. That lovely name fascinated me for years, as I listened to the tales that my parents told. People of long ago and very far away are really fairy-like to a child. In my maturity, when I returned to Ireland and searched for my roots, I put together dad's little vignettes, sought out the old homestead, and visited the church where I leafed through the worn faded pages of the register. And voilà—the marriage of Richard Sullivan and Jane Justice on November 10, 1861. The names of those who stood up with them. On other pages, the baptisms of their children, here in the church where they went to Mass. Of the six Sullivan children, one girl went to Australia, another died young, and the four boys to America, including my father, around 1901, to Lawrence, Massachusetts.

Ita Donahue, holding the pot of shamrocks that she brought back from one of her visits to Ireland. Her Irish linen dress came from Standun's, a store in County Galway.

A portrait of Lawrence Justice Sullivan, Ita Donahue's father, who immigrated to the United States around 1901.

Q: Why did they go to Lawrence?

B ecause Lawrence and Lowell [towns in northeastern Massachusetts] already had many people from their own area of Ireland. The cloth mills had hired Irish workers for some time. My dad worked as an examiner in the Wood Mill in Lawrence, and there he met my mother, who was a mender.

She had been born in County Limerick in 1882, near a village called Mountcollins, only about 15 miles away from my father's place. That was O'Connor land, and Mom was an O'Connor. Johannah O'Connor. She lived on a farm in Secont Glish, the youngest of nine. After Mom's parents died, she came to America. It was popular at that time for girls to do so. They felt secure in coming, because friends who arrived earlier would "claim" them. The "claim" assured their housing, care, and support until the newcomers found jobs. Irish girls were much in demand for mill work or housekeeping.

It was at this point that Sullivan met O'Connor, and their marriage took place in St. Mary's Shrine in Lawrence on Thanksgiving, 1908. I was born 12 years later, an only child.

Q: But your parents returned to Ireland for a while.

T hat's right. My father took a job in another mill, which closed. He had saved some money, and had never forgotten that road lined with flowers. So he and my mother sold their home and possessions, and gave away their treasured King Charles spaniel. The three of us packed up and went home to Ireland for what they thought was forever. That was in 1921. This picture with the donkey cart was taken in Secont Glish. That's me, I was just about one year old, with relatives and my mom in her raccoon coat and big beautiful hat, and her boots. I see in this a story.

And the story goes, Dad's intention was to purchase a place near Mountcollins. He found one out on the Listowel road on the way to Ballybunion, in County Kerry. It had a pub, store, and a fine house. Dad was going to buy that. He was an ambitious man. But it also had a caretaker who wished to stay.

This was a time filled with Ireland's worst troubles. From the uprising in 1914 into the early 20s, fighting to gain independence from England was at its height. Eamon De Valera, the Irish leader, sent envoys including Michael Collins to England to discuss a settlement. De Valera had told them not to sign anything without getting permission from him. But the English threatened them with ruthless attacks on Ireland the likes of which they had never seen before unless they signed a treaty immediately.

They signed the treaty in December 1921, and returned to Ireland, saying, "It was the best we could do." But De Valera and his followers said, "No. Ourselves alone!" Now, the fighting became brother against brother, hero against hero, and a civil war broke out that lasted until 1923. People still remember that time; there are De Valera men and Michael Collins men today!

Johannah O'Connor Sullivan stands by as one-year-old Ita, her daughter, holds the reins of a donkey cart in Secont Glish, Johannah's birthplace. The year was 1921, and the Sullivan family had just returned to Ireland for what they thought was forever. However, events forced them to move back to the United States.

A portrait of young Ita Donahue with her parents in Lawrence, Massachusetts, in the early 1920s.

Four generations of an Irish American family. On the couch, from left, Ita Donahue, her daughter Joan Donahue Griffin, and Joan's daughter Tara Griffin. The painting on the wall shows Donahue's mother on a visit to Ireland in 1970, when she was 89.

Dressed as St. Patrick, John Griffin, Ita Donahue's son-in-law, leads the annual St. Patrick's Day Parade in Lawrence.

To complete the tale, we were right there in the middle of it as my father was thinking about buying this place. In some eyes, he was considered "the Yank," as those who immigrated were called. He was told that whoever went through the town and took it, could be a friend of the caretaker and we could lose the place.

So we went down to Ballybunion, close to the beach and the castle, and lived for a year to wait for the settlement. One night, the railroad station went up in flames. One or the other side in the civil war had set it afire. Dad and Mom packed our suitcases that night and, with a farewell to Mountcollins and our people, we went back to America. That spoiled my nice Irish accent, but imagine the heartbreak it must have been for my parents.

We resettled in Lawrence. My dad did well for himself. He bought a house five doors down from the man who owned his company. I went through the local schools and enrolled in Emmanuel College, a Catholic women's college in Boston, in 1941. That was part of the very strong influence of my parents.

Q: They wanted you to get a college education.

But my father especially wanted me to be a teacher. It was not really my goal at first, and when I graduated, it was wartime, and I worked for General Electric and taught evening school. I was married during the war [World War II] and had my only child, Joan.

Today she's a parent liaison in the Lawrence school system and is married to John Griffin, himself a high school teacher. His forebears are from Dingletown in County Kerry, and were fast friends of Joan's grandparents. A young colleen also graces our household—their daughter Tara Kathleen, who's in her last year of high school, ready to go to college herself.

But yes, I did turn to teaching full time at the junior high school, and can honestly say that from the day I put my foot in a classroom, I never knew a boring day. If I had to do life again, I'd do exactly the same. I was given the assistant principalship of Woodbury Junior High here in Salem and later became the principal of Fisk Elementary. I was in the Salem system for 30 years before retirement.

After I retired, the children made this banner for me, with the Gaelic inscription, DIA IS MAIRE DHUIT. It is a greeting in Ireland, but it means "God and Mary be with you." Now, what happened was that a teacher went to the library to find a suitable Gaelic inscription. The librarians discussed it, and not knowing I would be the recipient, called me for advice!

Q: They knew that you were interested in Irish culture.

Yes, in everything Irish. People see us on the street, and they say oh, Ireland. We must project it! I was formerly the president of the Ladies Ancient Order of Hibernians division in this area. Joan and I are active in its affairs, and Tara is a member of the Young Hibernians Society. The Hibernians sponsor classes in Irish dancing, language, and culture. Irish step dancing is taught on Mondays, and while some of us can't get the heel up as high as the young people, we can all enjoy the special dances—the "Siege of Ennis" and the "Stack of Barley" when we have a real Irish dance at Hibernian Hall. My mother was a great step dancer herself and a violinist. There is great music in the Irish.

Tomorrrow's parade is a major highlight in the greater Lawrence area, for thousands view it. The Hibernians ran it for many years, but in 1983, they needed to put money into buying a clubhouse. So one year there wasn't a parade, and there was such an outcry that a new group of citizens gathered to keep the tradition alive. The St. Patrick's Parade Committee, including this household, will see the fruits of their labors tomorrow. The parade not only honors St. Patrick and Ireland, but excites an entire community as they don the green. Such a happy day for young and old!

Q: Tell us about the times you went back to Ireland as an adult, to find your roots.

I took my mother back twice, the first time in 1970. At age 89, my mother made her return after 50 years to a valley filled with old relations and school chums. Out came the fiddles and the accordions and there was many a set dance in the kitchens! We went again when she was 92, to the same reception. I have always felt bad I didn't get my Dad back, because he died in 1964, without seeing his road lined with flowers.

It was there I fully understood the derivation of my own name, Ita. When I was a child, Mom had told me it was an Irish saint's name. I remember somebody—another girl, about 12—said to me, "What's your name?" When I told her, she said, "Oh! I don't like that!" My mother never sat me down and told me the full story.

In Ireland in 1970, however, Mom pointed out a hill at the rear of the family farm called the Boolavita, which means in Irish "dairy of Ita." As a young girl centuries ago, St. Ita used to tend her goat on this hill, and so its name. At a nearby town, today called Kileedy (village of Ita), she, like St. Bridget, opened a school to train young girls for religious orders. The name Ita is heard occasionally in Ireland, very rarely in America, but I don't mind anymore, for it does have a special meaning for me now.

The certificate of Ita Marie O'Connor Sullivan Donahue's membership in the O'Connor Kerry Clan.

Ita Donahue is seated just to the left of the accordionist at this reunion of the O'Connor Clan in Ireland.

IRISH AMERICAN TIMELINE

1620s–1770s
Irish immigrants arrive in the English colonies of North America; many were indentured servants.

1681
The first Charles Carroll immigrates to Maryland after his estates in Ireland are confiscated by the English government.

1703
The English government of Ireland begins to use "transportation" to the American colonies as punishment for rebels and criminals.

1759
First St. Patrick's Day Parade in New York City.

1776
Charles Carroll of Carrollton (third family member of that name) becomes the only Catholic to sign the Declaration of Independence. Many Irish, both Catholic and Protestant, fight on the American side in the Revolution.

1783–1845
About 400,000 Irish immigrants arrive in the United States—more than any other ethnic group in that era.

1789
John Carroll, first American Catholic bishop, founds the school that later becomes Georgetown University.

1794
Irish-born John Barry, naval hero of the American Revolution, is named to head the U.S. Navy.

1836
The U.S. branch of the Ancient Order of Hibernians is founded in New York City.

1844
Anti-Catholic riots in Philadelphia; part of antiforeign sentiment that leads to the formation of the American party, called the Know-Nothings.

1845
A blight begins to destroy the potato crop of Ireland and continues into the 1850s. The resulting Great Famine causes the death of about 1 million Irish citizens.

1845–1860
Almost 2 million emigrants leave Ireland. Most are headed for the United States.

1849
California gold rush brings a wave of Irish immigrants to California.

1855
Castle Garden in New York Harbor opens as a landing station for immigrants.

1858
Former Irish rebels in Ireland and the United States found the Fenian Brotherhood, an organization dedicated to winning Irish independence.

1879
Terence Powderly becomes head of the Knights of Labor.

1886
Hugh O'Brien becomes the first Irish American mayor of Boston. In the last half of the 19th century, Irish Americans in many cities become active in politics.

1892
Ellis Island replaces Castle Garden as the main immigrant landing station in the United States. First person to pass through is Irish immigrant Annie Moore.

1928
Alfred E. Smith loses his bid to become the first Irish American Catholic president of the United States.

1936
Eugene O'Neill wins the Nobel Prize for Literature.

1960
John F. Kennedy is elected President of the United States.

1981
Sandra Day O'Connor becomes the first woman justice of the Supreme Court.

FURTHER READING

General Accounts of Irish American History

Bradley, Ann Kathleen. *History of the Irish in America.* Secaucus, N.J.: Chartwell Books, 1986.

Greeley, Andrew. *The Irish Americans: The Rise to Money and Power.* New York: Warner, 1981.

Griffin, William D. *A Portrait of the Irish in America.* New York: Scribners, 1981.

Miller, Kerby A. *Emigrants and Exiles: Ireland and the Irish Exodus to North America.* New York: Oxford University Press, 1985.

Miller, Kerby, and Paul Wagner. *Out of Ireland: The Story of Irish Emigration to America.* Washington, D.C.: Elliott and Clark, 1994.

Shannon, William V. *The American Irish: A Political and Social Portrait.* 2nd ed. Amherst: University of Massachusetts Press, 1989.

Specific Aspects of Irish American Life

Adams, William Forbes. *Ireland and Irish Emigration to the New World from 1815 to the Famine.* 1932. Reprint. New York: Russell & Russell, 1967.

Birmingham, Stephen. *Real Lace: America's Irish Rich.* New York: Harper & Row, 1973.

Cooper, Brian E., ed. *The Irish-American Almanac and Green Pages.* New York: Pembroke Press, 1986.

Diner, Hasia R. *Erin's Daughters in America: Irish Immigrant Women in the Nineteenth Century.* Baltimore: Johns Hopkins University Press, 1983.

Flannery, John Brendan. *The Irish Texans.* San Antonio: Institute of Texan Cultures, 1980.

McGee, John Whalen. *The Passing of the Gael.* Grand Rapids, Mich.: Wolverine Printing, 1975.

Mitchell, Brian C. *The Paddy Camps: The Irish of Lowell, 1821–1861.* Urbana: University of Illinois Press, 1988.

O'Carroll, Ide. *Models for Movers: Irish Women's Emigration to America.* Dublin: Attic Press, 1990.

Ridge, John T. *Erin's Sons in America: The Ancient Order of Hibernians.* New York: AOH 150th Anniversary Committee, 1986.

———. *The St. Patrick's Day Parade in New York.* New York: St. Patrick's Day Committee, 1988.

Sarbaugh, Timothy J., and James P. Walsh, eds. *The Irish in the West.* Manhattan, Kans.: Sunflower University Press, 1993.

Schrier, Arnold. *Ireland and the American Emigration 1850–1900.* Minneapolis: University of Minnesota Press, 1958.

Personal Accounts of Irish American Life

Butte Historical Society. *Looking Back From the Hill: Recollections of Butte People.* Butte, Mont.: Butte Historical Society, 1982.

Byrne, Jane. *My Chicago.* New York: Norton, 1992.

Byrnes, Thomas. *My Angel's Name is Fred: Tales of Growing Up Catholic.* San Francisco: Harper & Row, 1987.

Cahill, Kevin M., ed. *The American Irish Revival.* Port Washington, N.Y.: Associated Faculty Press, 1984.

Cather, Thomas. *Thomas Cather's Journal of a Voyage to America in 1836.* Emmaus, Pa.: Rodale Press, 1955.

Cohen, Stephen, ed. *America the Dream of My Life.* New Brunswick, N.J.: Rutgers University Press, 1990.

Curley, James Michael. *I'd Do It Again.* Englewood Cliffs, N.J.: Prentice-Hall, 1957.

Donohue, Michael. *Starting Off From Dead End.* New York: Community Documentation Workshop, St. Mark's-in-the-Bowery Church, 1980.

Hackett, Francis. *American Rainbow.* New York: Liveright, 1971.

Kalergis, Mary Motley. *Home of the Brave.* New York: Dutton, 1989.

Kearns, Kevin C. *Dublin Street, Life and Lore.* Dublin: Glendale Press, 1991.

Kelly, Mary Ann. *My Old Kentucky Home, Good-night.* Hicksville, N.Y.: Exposition Press, 1979.

Kisseloff, Jeff. *You Must Remember This: An Oral History of Manhattan from the 1890s to World War II.* San Diego: Harcourt Brace Jovanovich, 1989.

Lawlor, David S. *The Life and Struggles of an Irish Boy in America.* Newton, Mass.: Carroll, 1936.

Morrison, Joan, and Charlotte Fox Zabusky. *American Mosaic.* Pittsburgh: University of Pittsburgh Press, 1980.

O'Connell, William Cardinal. *Recollections of Seventy Years.* Boston: Houghton Mifflin, 1934.

O'Donovan Rossa, Diarmuid. *Rossa's Recollections 1838–1898.* 1898. Reprint. Shannon, Ireland: Irish University Press, 1972.

O'Dwyer, Paul. *Counsel for the Defense.* New York: Simon & Schuster, 1979.

Wright, Giles, Howard L. Green, and Lee R. Parks, eds. *The New Jersey Ethnic Life Series.* Trenton: New Jersey Historical Commission, 1986.

TEXT CREDITS

Main Text

p. 12, top: Cecil Woodham-Smith, *The Great Hunger* (New York: Harper & Row, 1962), 91.

p. 12, bottom: Diarmuid O'Donovan Rossa, *Rossa's Recollections 1838–1898* (1898; reprint, Shannon, Ireland: Irish University Press, 1972), 108-9.

p. 13, top: *The Times* (London), December 24, 1846.

p. 13, bottom: Stephen Birmingham, *Real Lace* (New York: Harper & Row, 1973), 13.

p. 14, top: "Memoirs of Rosalie B. Hart Priour," typescript in the Library of the Institute of Texan Cultures at San Antonio.

p. 14, bottom: *Illustrated London News*, December 22, 1849.

p. 15, top: From *Counsel for the Defense*, by Paul O'Dwyer, 44-45. Copyright © 1979 by Paul O'Dwyer. Reprinted by permission of Simon & Schuster, Inc.

p. 15, bottom: June Namias, *First Generation* (Boston: Beacon Press, 1978), 14-16.

p. 16: Hamilton Holt, ed., *The Life Stories of Undistinguished Americans as Told by Themselves* (1906; reprint, New York: Routledge, 1990), 88-89.

pp. 17, 18: Reprinted from *American Mosaic: The Immigrant Experience in the Words of Those Who Lived It*, by Joan Morrison and Charlotte Fox Zabusky, by permission of the University of Pittsburgh Press. © 1980, 1993 by Joan Morrison and Charlotte Fox Zabusky.

p. 19, top: Reprinted from *American Rainbow, Early Reminiscences*, by Francis Hackett, by permission of the Liveright Publishing Corporation. Copyright © 1971 by Liveright Publishing Corporation.

p. 19, bottom: Giles R. Wright, comp., *Looking Back*, New Jersey Ethnic Life Series no. 10 (Trenton: New Jersey Historical Commission, 1986), 28-30.

p. 26, top: Holt, *The Life Stories*, 89-90.

p. 26, bottom: Arnold Schrier, *Ireland and the American Emigration 1850–1900* (Minneapolis: University of Minnesota Press, 1958), 40-41.

p. 27, first: Schrier, *Ireland and the American Emigration*, 24.

p. 27, second: Schrier, *Ireland and the American Emigration*, 24.

p. 27, third: Reprinted from *American Rainbow, Early Reminiscences*, by Francis Hackett, by permission of the Liveright Publishing Corporation. Copyright © 1971 by Liveright Publishing Corporation.

p. 27, fourth: Maurice O'Sullivan, *Twenty Years A-Growing* (New York: Viking Press, 1933), 219.

p. 28: Rossa, *Rossa's Recollections*, 142.

p. 29, first: Schrier, *Ireland and the American Emigration*, 90.

p. 29, second: *Correspondence on the Treatment of the Passengers on Board the Emigrant Ship "Washington,"* House Committee, 32nd Congress, 1851, vol. 40, p. 434.

p. 29, third: *British Government Report of the Select Committee on the Passenger's Act, 1851*, British Parliamentary Paper, 1852, vol. 18.

p. 29, fourth: David S. Lawlor, *The Life and Struggles of an Irish Boy in America* (Newton, Mass.: Carroll, 1936), 18, 22-23.

p. 30: From *Counsel for the Defense*, by Paul O'Dwyer, 50-51. Copyright © 1979 by Paul O'Dwyer. Reprinted by permission of Simon & Schuster, Inc.

p. 31, top: Schrier, *Ireland and the American Emigration*, 89.

p. 31, bottom: Kitty Fitzgerald Donovan, interview with Dorothy Hoobler, fall 1993.

p. 32, top: Thomas Cather, *Journal of a Voyage to America in 1836* (Emmaus, Pa.: Rodale Press, 1955), 14-15.

p. 32, bottom: *Correspondence on the Treatment of the Passengers on Board the Emigrant Ship "Washington,"* 434.

p. 33: Richard Howard Brown, "I Am of Ireland," American Irish Historical Society *Recorder* (1975): 124-125.

p. 34, first: Holt, *The Life Stories*, 89-90.

p. 34, second: From McConnell family letters in the Toronto Public Library manuscripts collection.

p. 34, third: Woodham-Smith, *The Great Hunger*, 226

p. 35, top: Lawlor, *Life and Struggles*, 22-23.

p. 35, bottom: Kitty Fitzgerald Donovan, interview with Dorothy Hoobler, fall 1993.

p. 36, top: Quoted in William D. Griffin, *The Book of Irish Americans* (New York: Random House, 1990), 23.

p. 36, bottom: Quoted in William D. Griffin, *The Irish in America, 550-1972* (Dobbs Ferry, N.Y.: Oceana Publications, 1973), 69.

p. 37, top: Quoted in Griffin, *The Irish in America*, 61-63.

p. 37, bottom: Quoted in Leslie Allen, *Liberty: The Statue and the American Dream* (New York: Summit Books, 1985), 55.

p. 40, top: "Memoirs of Rosalie B. Hart Priour," typescript in the Library of the Institute of Texan Cultures at San Antonio.

p. 40, bottom: John Francis Maguire, *The Irish in America* (1868; reprint, New York: Arno Press, 1969), 190.

p. 41, first: Stephen Byrne, *Irish Emigration to the United States* (1873; reprint, New York: Arno Press, 1969), 29-30.

p. 41, second: Byrne, *Irish Emigration to the United States*, 30.

p. 41, third: Ellis Island Oral History Project interview on February 10, 1986.

p. 42, top: Ellis Island Oral History Project interview on March 22, 1991.

p. 42, bottom: Giles R. Wright, comp., *The Journey from Home*, New Jersey Ethnic Life Series no. 2 (Trenton: New Jersey Historical Commission, 1986) 22.

p. 43, top and bottom: Thomas Kessner and Betty Boyd Caroli, *Today's Immigrants: Their Stories* (New York: Oxford University Press, 1982), 88.

p. 44: Cather, *Journal of a Voyage to America*, 18-19.

p. 45, top: Schrier, *Ireland and the American Emigration*, 31.

p. 45, bottom: Florence Elizabeth Gibson, *The Attitudes of the New York Irish toward State and National Affairs, 1848–1892* (New York: Columbia University Press, 1954), 15.

p. 46, first: William Alfred, "An Irish Integrity," quoted in Thomas W. Wheeler, ed., *The Immigrant Experience: The Anguish of Becoming American* (New York: Dial Press, 1971), 20-21.

p. 46, second: Ivan Chermayeff, Fred Wasserman, and Mary J. Shapiro, *Ellis Island: An Illustrated History of the Immigrant Experience* (New York: Macmillan, 1991) 43.

p. 46, third: Reprinted from *American Rainbow, Early Reminiscences*, by Francis Hackett, by permission of the Liveright Publishing Corporation. Copyright © 1971 by Liveright Publishing Corporation.

p. 47: Reprinted from *American Mosaic: The Immigrant Experience in the Words of Those Who Lived It*, by Joan Morrison and Charlotte Fox Zabusky, by permission of the University of Pittsburgh Press. © 1980, 1993 by Joan Morrison and Charlotte Fox Zabusky.

p. 48, top: *Journal of the American Irish Historical Society*, vol. 12, 1912.

p. 48, bottom: *Report of Select Committee on Colonization from Ireland*, House Committee, 1849, vol. 11.

p. 49, first: Schrier, *Ireland and the American Emigration*, 25.

p. 49, second: Mary Ann Kelly, *My Old Kentucky Home, Good-night* (Hicksville, N.Y.: Exposition Press, 1979) 104.

p. 49, third: Byrne, *Irish Emigration*, 26.

p. 49, fourth: Reprinted from *American Mosaic: The Immigrant Experience in the Words of Those Who Lived It*, by Joan Morrison and Charlotte Fox Zabusky, by permission of the University of Pittsburgh Press. © 1980, 1993 by Joan Morrison and Charlotte Fox Zabusky.

p. 54, first: Roger Daniels, *Coming to America* (New York: HarperCollins, 1991), 268.

p. 54, second: John Tracy Ellis, ed., *Documents of American Catholic History* (Chicago: Regnery, 1967), 114.

p. 54, third: Brian C. Mitchell, *The Paddy Camps: The Irish of Lowell 1821–1861* (Urbana: University of Illinois Press, 1988), 139-40.

p. 55: Mitchell, *The Paddy Camps*, 137-38.

p. 56, top: William Cardinal O'Connell, *Recollections of Seventy Years* (Boston: Houghton Mifflin, 1934), 5-6.

p. 56, bottom: From *I'd Do It Again*, by James Michael Curley © 1957, 1985. Used by permission of the publisher, Prentice Hall/A Division of Simon & Schuster, Englewood Cliffs, N.J., 9-10.

p. 57, top: Howard L. Green and Lee R. Parks, comps., *What Is Ethnicity?*, New Jersey Ethnic Life Series no. 8, (Trenton: New Jersey Historical Commission, 1987), 19.

p. 57, bottom: Schrier, *Ireland and the American Emigration*, 36.

p. 58, top: Tyrone Power, "Irish Workmen Build a Canal near New Orleans," in Rhoda Hoff, *America's Immigrants* (New York: Henry Z. Walck, 1967), 22.

p. 58, bottom: O'Connell, *Recollections*, 191.

p. 59, top: Lawlor, *The Life and Struggles*, 31.

p. 59, bottom: Mary Harris Jones, *Autobiography of Mother Jones* (Chicago: C. H. Kerr, 1925), 121-22.

p. 60, top: From *America the Dream of My Life*, edited by David Steven Cohen, copyright © 1990 by Rutgers, the State University. Reprinted by permission of Rutgers University Press, 41-42.

p. 60, bottom: Michael Donohue, *Starting Off From Dead End* (New York: Community Documentation Workshop of St. Mark's Church in-the-Bowery, 1980), 15-16, 24-26.

p. 61: Excerpts from *You Must Remember This: An Oral History of Manhattan from the 1890s to World War II*, copyright © 1989 by Jeff Kisseloff, reprinted by permission of Harcourt Brace & Company.

p. 62, top: From *Counsel for the Defense*, by Paul O'Dwyer, 57-59. Copyright © 1979 by Paul O'Dwyer. Reprinted by permission of Simon & Schuster, Inc.

p. 62, bottom: Reprinted from *American Mosaic: The Immigrant Experience in the Words of Those Who Lived It*, by Joan Morrison and Charlotte Fox Zabusky, by permission of the University of Pittsburgh Press. © 1980, 1993 by Joan Morrison and Charlotte Fox Zabusky.

p. 63: Reprinted from *American Mosaic: The Immigrant Experience in the Words of Those Who Lived It*, by Joan Morrison and Charlotte Fox Zabusky, by permission of the University of Pittsburgh Press. © 1980, 1993 by Joan Morrison and Charlotte Fox Zabusky.

p. 64: Holt, *The Life Stories*, 90-92.

p. 65: Ann Banks, *First Person America* (New York: Knopf, 1980) 44-45.

p. 66: Reprinted from *American Mosaic: The Immigrant Experience in the Words of Those Who Lived It*, by Joan Morrison and Charlotte Fox Zabusky, by permission of the University of Pittsburgh Press. © 1980, 1993 by Joan Morrison and Charlotte Fox Zabusky.

p. 67: Reprinted from *American Mosaic: The Immigrant Experience in the Words of Those Who Lived It*, by Joan Morrison and Charlotte Fox Zabusky, by permission of the University of Pittsburgh Press. © 1980, 1993 by Joan Morrison and Charlotte Fox Zabusky.

p. 68, top: Lawlor, *The Life and Struggles*, 32.

p. 68, bottom: Jones, *Autobiography*, 13-14,

p. 69: Jones, *Autobiography*, 84-87.

p. 71, first: Quoted in Griffin, *The Book of Irish Americans*, 120-21.

p. 71, second and third: "Memoirs of Rosalie B. Hart Priour," typescript in the Library of the Institute of Texan Cultures at San Antonio.

p. 72: *Looking Back from the Hill: Recollections of Butte People* (Butte, Mont.: Butte Historical Society, 1982), 66, 68.

p. 74, top: From *I'd Do It Again*, by James Michael Curley © 1957, 1985. Used by permission of the publisher, Prentice Hall/A Division of Simon & Schuster, Englewood Cliffs, N.J., 13-14, 18.

p. 74, bottom: William L. Riordon, ed., *Plunkitt of Tammany Hall* (New York: McClure, Phillips, 1905), 51-52.

p. 75, first: Riordon, *Plunkitt of Tammany Hall*, 58-59.

p. 75, second: Quoted in Carl Wittke, *The Irish in America* (Baton Rouge: University of Louisiana Press, 1956), 154.

p. 75, third: Finley Peter Dunne, *Mr. Dooley in the Hearts of His Countrymen* (Boston: Small, Maynard, 1899), 35.

p. 76, top: Excerpts from *You Must Remember This: An Oral History of Manhattan from the 1890s to World War II*, copyright © 1989 by Jeff Kisseloff, reprinted by permission of Harcourt Brace & Company.

p. 76, bottom: Lloyd Wendt and Herman Kogan, *Bosses in Lusty Chicago* (Bloomington: Indiana University Press, 1976), v-vi.

p. 77, top: Donohue, *Starting Off From Dead End*, 19-20.

p. 77, bottom: Alfred E. Smith, "Defense of Catholics in Public Office," quoted in *Annals of America*, vol. 14 (Chicago: Encyclopedia Britannica, 1968), 536-37.

p. 82, top: Alfred E. Smith, *Up to Now* (New York: Viking, 1929), 25-27.

p. 82, bottom: From *I'd Do It Again*, by James Michael Curley © 1957, 1985. Used by permission of the publisher, Prentice Hall/A Division of Simon & Schuster, Englewood Cliffs, N.J., 34-35.

p. 83: *Looking Back from the Hill: Recollections of Butte People*, 67-68, 70.

p. 84: Michael Doyle, manuscript at the Balch Institute for Ethnic Studies Library, Philadelphia, 7-8.

p. 85, first: Excerpts from *You Must Remember This: An Oral History of Manhattan from the 1890s to World War II*, copyright © 1989 by Jeff Kisseloff, reprinted by permission of Harcourt Brace & Company.

p. 85, second: Excerpts from *You Must Remember This: An Oral History of Manhattan from the 1890s to World War II*, copyright © 1989 by Jeff Kisseloff, reprinted by permission of Harcourt Brace & Company.

p. 85, third: From *America the Dream of My Life*, edited by David Steven Cohen, copyright © 1990 by Rutgers, the State University. Reprinted by permission of Rutgers University Press, 36.

p. 86, top: Mitchell, *The Paddy Camps*, 103-4.

p. 86, bottom: Margaret Sanger, *An Autobiography* (New York: Norton, 1938), 12-13.

p. 87: Kelly, *My Old Kentucky Home*, 109.

p. 88, top: Lawlor, *The Life and Struggles*, 42-44.

p. 88, bottom: *Looking Back from the Hill: Recollections of Butte People*, 66.

p. 89: Buster Keaton and Charles Samuels. *My Wonderful World of Slapstick* (Garden City, N.Y.: Doubeday, 1960), 12-14.

p. 90: Banks, *First Person America*, 171-72.

p. 92, top: Namias, *First Generation*, 19-20.

p. 92, bottom: Robert C. Hayden, ed., *A Dialogue with the Past* (Newton, Mass.: Education Development Center, 1979), 19-20.

p. 93: Alameda King, interview with Dorothy Hoobler, spring 1994.

p. 94, top: Jane Byrne, *My Chicago* (New York: Norton, 1992) 18-19.

p. 94, bottom: Selected excerpts from *My Angel's Name Is Fred* by Thomas Byrnes. Copyright © 1987 by Thomas Byrnes. Reprinted by permission of HarperCollins Publishers, Inc. Pages 112-13, 116-17.

p. 96, top: O'Connell, *Recollections*, 47.

p. 96, middle: Excerpts from *You Must Remember This: An Oral History of Manhattan from the 1890s to World War II*, copyright © 1989 by Jeff Kisseloff, reprinted by permission of Harcourt Brace & Company.

p. 96, bottom: Excerpts from *You Must Remember This: An Oral History of Manhattan from the 1890s to World War II*, copyright © 1989 by Jeff Kisseloff, reprinted by permission of Harcourt Brace & Company.

p. 97: Excerpts from *You Must Remember This: An Oral History of Manhattan from the 1890s to World War II*, copyright © 1989 by Jeff Kisseloff, reprinted by permission of Harcourt Brace & Company.

p. 98: Griffin, *The Irish in America, 550–1972*, 301-4.

p. 99, top: John T. Ridge, *Erin's Sons in America: The Ancient Order of Hibernians* (New York: Ancient Order of Hibernians 150th Anniversary Committee, 1986), 10.

p. 99, bottom: John T. Ridge, *Sligo in New York: The Irish from County Sligo: 1849–1991* (New York: County Sligo Social and Benevolent Association, 1991), 51.

p. 100, first: From *I'd Do It Again*, by James Michael Curley © 1957, 1985. Used by permission of the publisher, Prentice Hall/A Division of Simon & Schuster, Englewood Cliffs, N.J., 45-46.

p. 100, second: Griffin, *The Book of Irish Americans*, 137.

p. 100, third: Lawrence J. McCaffrey, *Textures of Irish America* (Syracuse, N.Y.: Syracuse University Press, 1992), 138.

p. 100, fourth: Rossa, *Rossa's Recollections*, 199.

p. 101, top: Joseph I. C. Clarke, *My Life and Memories* (New York: Dodd, Mead, 1925), 39, 44.

p. 101, bottom: William V. Shannon, *The American Irish: A Political and Social Portrait* (Amherst: University of Massachusetts Press, 1989), 133.

p. 106, top: *New York Times*, September 13, 1960.

p. 106, bottom: Jimmy Cannon, *Nobody Asked Me, But.... The World of Jimmy Cannon* (New York: Henry Holt, 1978).

p. 107: Mary Higgins Clark, "My Wild Irish Mother," in *The American Irish Revival—A Decade of The Recorder 1974–1983*, edited by Kevin M. Cahill (Port Washington, N.Y.: Associated Faculty Press, 1984), 689-99. Copyright © 1977 by Mary Higgins Clark. Reprinted by permission of McIntosh and Otis, Inc.

p. 109: Wright, *Looking Back*, 32-34.

p. 110: Kessner and Caroli, *Today's Immigrants*, 144-45.

p. 111: Mary Motley Kalergis, *Home of the Brave* (New York: Dutton, 1989), unpaged.

p. 112, top: Kevin C. Kearns, *Dublin Street Life and Lore: An Oral History* (Dublin: Glendale, 1991), 228.

p. 112, bottom: Kessner and Caroli, *Today's Immigrants*, 146.

p. 113: Ide O'Carroll, *Models for Movers: Irish Women's Emigration to America* (Dublin: Attic Press, 1990), 118-20.

p. 114: O'Connell, *Recollections*, 42-43.

p. 115, top: *Looking Back from the Hill: Recollections of Butte People*, 67-68.

p. 115, bottom: Tom Bernardin, *The Ellis Island Immigrant Cookbook*, copyright © 1991 by Tom Bernardin, New York, NY 10010, 176-77.

p. 116, top: Kitty Fitzgerald Donovan, interview with Dorothy Hoobler, fall 1993.

p. 116, bottom: Ide O'Carroll, *Models for Movers: Irish Women's Emigration to America* (Dublin: Attic Press, 1990), 129-31.

p. 117: Kevin M. Cahill, "The Descendants of the High Kings of Ireland," in *The American Irish Revival—A Decade of The Recorder 1974–1983*, edited by Kevin M. Cahill (Port Washington, N.Y.: Associated Faculty Press, 1984), 804-5.

Sidebars

p. 13: Charles Gavan Duffy, *Four Years of Irish History, 1845–1849* (New York: 1882)

p. 14: Reprinted from *American Mosaic: The Immigrant Experience in the Words of Those Who Lived It*, by Joan Morrison and Charlotte Fox Zabusky, by permission of the University of Pittsburgh Press. © 1980, 1993 by Joan Morrison and Charlotte Fox Zabusky.

p. 16: Carter G. Woodson, ed., *The Mind of the Negro As Reflected in Letters Written During the Crisis 1800–1860* (New York: Negro Universities Press, 1969), 410-11.

p. 21: Hasia R. Diner, *Erin's Daughters in America: Irish Immigrant Women in the Nineteenth Century* (Baltimore: Johns Hopkins University Press, 1983), 17-18.

p. 26: Giles Wright, *Arrival and Settlement in a New Place*, New Jersey Ethnic Life Series no. 3 (Trenton: New Jersey Historical Commission, 1986), 29.

p. 27: Reprinted from *American Rainbow, Early Reminiscences*, by Francis Hackett, by permission of the Liveright Publishing Corporation. Copyright © 1971 by Liveright Publishing Corporation.

p. 28: Arnold Schrier, *Ireland and the American Emigration 1850-1900* (Minneapolis: University of Minnesota Press, 1958), 89.

p. 29: Stephen Byrne, *Irish Emigration to the United States* (New York: The Catholic Publication Society, 1873), 28.

p. 30: Timothy J. Sarbaugh and James P. Walsh, eds., *The Irish in the West* (Manhattan, Kans.: Sunflower University Press, 1993), 63.

p. 32: Ivan Chermayeff, Fred Wasserman, and Mary J. Shapiro, *Ellis Island: An Illustrated History of the Immigrant Experience* (New York: Macmillan, 1991), 168.

p. 33: From *Counsel for the Defense*, by Paul O'Dwyer, 56-57. Copyright © 1979 by Paul O'Dwyer. Reprinted by permission of Simon & Schuster, Inc.

p. 35: Ronald Takaki, *A Different Mirror: A History of Multicultural America* (Boston: Little, Brown, 1993) 145.

p. 41: Diner, *Erin's Daughters*, 38.

p. 43: James P. Shenton, and Gene Brown, eds., *Ethnic Groups in American Life* (New York: Arno Press, 1978), 52.

p. 44: Kitty Fitzgerald Donovan, interview with Dorothy Hoobler, fall 1993.

p. 45: Wright, *Arrival and Settlement in a New Place*, 18.

p. 46: Giles Wright, comp., *The Journey from Home*, New Jersey Ethnic Life Series no. 2, (Trenton: New Jersey Historical Commission, 1986), 39.

p. 48: Terry Coleman, *Going to America* (New York: Pantheon, 1972), 25-26.

p. 54: *Annals of America*, vol. 7 (Chicago: Encyclopedia Britannica, 1968), 421.

p. 55: Coleman, *Going to America*, 228.

p. 58: Maldwyn Allen Jones, *American Immigration* (Chicago: University of Chicago Press, 1960), 32.

p. 60: Diner, *Erin's Daughters*, 78.

p. 61: Diner, *Erin's Daughters*, 71.

p. 65: Takaki, *A Different Mirror*, 158.

p. 67: From *I'd Do It Again*, by James Michael Curley © 1957, 1985. Used by permission of the publisher, Prentice Hall/A Division of Simon & Schuster, Englewood Cliffs, N.J., 36.

p. 68: Elizabeth Gurley Flynn, *The Rebel Girl: an Autobiography* (New York: International Publishers, 1955, 1973), 135.

p. 74: From *I'd Do It Again*, by James Michael Curley © 1957, 1985. Used by permission of the publisher, Prentice Hall/A Division of Simon & Schuster, Englewood Cliffs, N.J., 49.

p. 77: William D. Griffin, *The Book of Irish Americans* (New York: Random House, 1990), 107.

p. 83: Brian C. Mitchell, *The Paddy Camps: The Irish of Lowell 1821–1861* (Urbana: University of Illinois Press, 1988), 26.

p. 85: Isaac A. Hourwich, *Immigration and Labor* (New York: Putnam's, 1912)

p. 87: Diner, *Erin's Daughters*, 69.

p. 89: Diner, *Erin's Daughters*, 68.

p. 90: R. A. Burchell, *The San Francisco Irish 1848–1880* (Berkeley: University of California Press, 1980), 77.

p. 95: *Donahoe's* magazine, November 1890, 346.

p. 96: Lawrence J. McCaffrey, *Textures of Irish America* (Syracuse: Syracuse University Press, 1992), 194.

p. 101: Elizabeth Gurley Flynn, *I Speak My Own Piece* (New York: Masses & Mainstream, 1955), 13.

p. 115: Tom Bernardin, *The Ellis Island Immigrant Cookbook*, copyright © 1991 by Tom Bernardin, New York, NY 10010, 177.

PICTURE CREDITS

Alaska State Library, L. R. Zacharias Collection (#PCA 178): 117 bottom; Alice Austen, Staten Island Historical Society Collection: 45; courtesy of Ancient Order of Hibernians, Lawrence, Massachusetts, 54; courtesy of Archives, Archdiocese of Boston: 93 top, 94 top; Archives, Catholic Diocese of Salt Lake City: 89, 96; the Balch Institute for Ethnic Studies Library: cover (McGovern Family Photographs), 8 (Mulhern Family Photographs), 26 (Balch Reproductions), 40 bottom (Lucas Family Photographs), 41 (Center for Immigration Research at the Balch Institute—Manifest, July 26, 1861), 47 bottom (Balch Reproductions), 117 top (Balch Institute Postcards); Bancroft Library: 75; Bettmann Archive: 19 bottom, 20 bottom, 80; Boston Gas Archives, John J. Burns Library, Boston College: 50; courtesy of the Boston Public Library, Prints Department: 74; courtesy Bostonian Society, Old State House: 65, 90; Brown Brothers, 33; California Museum of Photography, Keystone-Mast Collection, University of California, Riverside: 42; courtesy of the Carroll family: 17 bottom, 20 top, 25, 92; Chicago Historical Society (ICHi-10926): 76; Cleveland's Irish Cultural Festival: 113; Culver Pictures: 38; Denver Public Library, Western History Department: 70 top; courtesy Ita Donahue: 118 bottom, 119, 120, 121; Dublin University Department of Irish Folklore: 14; courtesy George Eastman House: 67; courtesy of Hagley Museum and Library: 32 top; Harper's Magazine, 57; Ellen Hoobler: 116; Thomas Hoobler: 118 top; Idaho State Historical Society: 5 (#70-101.1), 73 top left (#1037-21), 73 bottom (#552-A F754.B62), 104 (#247-A F754.B62); Immigrant City Archives, Lawrence, Massachusetts: 53, 60, 61, 83, 84 top, 86 bottom, 94 bottom, 95, 100, 105 (*Eagle-Tribune* collection); Institute of Texan Cultures, San Antonio, Texas: 29, 70 bottom, 86 top; *Irish-American Almanac*, 30 left; John F. Kennedy Library, Boston, Mass.: 6 top and middle, 106; courtesy of Lane County Historical Museum: 91 bottom; Library of Congress: 10, 17 top, 22, 30 right, 44, 55, 64 bottom, 68, 69, 78, 82 bottom, 84 bottom, 97 bottom, 101; Thomas McAllister, Museum of the City of New York: 49 top; Metropolitan Museum of Art, gift of David A. Schulte, 1928 (28.127.3): 109; Minnesota Historical Society: 97 top (William Sharkey), 87 (Randolph R. Johnson), 98 bottom; Museum of the City of New York, the Byron Collection: 28, 34 top, 34 bottom, 35, 59, 77 left, 82 top; Museum of the City of New York, Gift of Mrs. Robert M. Littlejohn: 40 top; National Archives: 37 top, 52, 62; by courtesy of the National Park Service, Ellis Island Immigrant Museum: 31 top, 37 bottom, 43 (K. Daley); by courtesy of the National Park Service, Statue of Liberty National Monument: 24; collection of the New-York Historical Society: 31 bottom, 46, 49 bottom, 88 bottom; New York Public Library Picture Collection: 15, 18, 85, 91 top, 99; New York State Library: 27, 32 bottom, 48; Mulcahy-Noel family: 98 top; Joanne O'Brien/Format/Impact Visuals: 110; Office of Congressman Joseph P. Kennedy II: 6 bottom left and right, 7; Joseph J. Pennell Collection, Kansas Collection, Kansas State Library: 71; Pennsylvania Historical and Museum Commission, Anthracite Museum Complex: 58; Antonio Perez, 102; Pollard Memorial Library, 66; Schlesinger Library, Radcliffe College: 16, 19 top, 111, 112; Sharlot Hall Museum Library/Archives, Prescott, Arizona: 73 top right; Smithsonian Institution: 56, 77 right; State Historical Society of Wisconsin, Charles J. Van Schaick: 64 top [WHi (V22) 1387]; the collection of the Supreme Court of the United States: 107; courtesy of Loretto Dennis Szucs: 63 top, 81, 88 top, 93 bottom; courtesy of Zirkle Thigpen: frontispiece; Ulster Museum: 13; UPI/Bettmann: 11, 21, 108; World Museum of Mining, Butte, Montana: 63 bottom.

INDEX

ACKNOWLEDGMENTS

We owe a debt of gratitude to John Concannon, national historian of the Ancient Order of Hibernians; Loretto Dennis Szucs, editor of the *Federation of Genealogical Societies Forum* and acquisitions editor of *Ancestry;* and Dr. Charles M. Hogan, for sharing their knowledge of Irish American history and giving us advice on how to pursue our research. Ken Skulski of the Immigrant City Archives made crucial contributions to this book with his encouragement, knowledge of immigrant history, contributions of pictures, and by introducing us to Ita Donahue. Tom Bernardin generously allowed us to quote from *The Ellis Island Immigrant Cookbook.* Our personal thanks to Paul McCarthy and his mother, Mary McCarthy; Loretto Dennis Szucs; Kitty Fitzgerald Donovan; and Mel Noel for contributing their family photographs, papers, and memories.

We also wish to acknowledge the help of John B. Atteberry of the Boston College Library; Philip Bergen of the Bostonian Society; Nicolette Bromberg and Andy Kraushaar of the State Historical Society of Wisconsin; Diane Bruce of the Institute of Texan Cultures; Angela Carter of the New York Irish History Roundtable and Irish Books and Graphics; Marian Casey; Canon Lloyd S. Casson of St. Mark's Church in-the-Bowery; John Corrigan; Ellen Crain of the Butte Historical Society; Tara Deal and Nancy Toff, our editors at Oxford University Press; Joseph Dolan, chairman of the Board of Trustees of the Irish American Heritage Museum; Carlotta de Fillo of the Staten Island Historical Society; Eileen Flanagan of the Chicago Historical Society; Thomas J. Fleming; Jim Francis of the New-York Historical Society; Marie-Hélène Gold and Wendy Thomas of the Schlesinger Library, Radcliffe College; Jean Hargrave of the New York State Library; Sinclair Hitchings and Aaron Schmidt of the Boston Public Library; Al Hooper of the World Museum of Mining; Thomas Horan of the American Irish Historical Society; Debra Hughes of the Hagley Museum and Library; Elizabeth P. Jacox of the Idaho State Historical Society; Dale Johnson of the Mansfield Library, University of Montana; Robert Johnson-Lally, archivist of the Archdiocese of Boston; Chester J. Kulesa, curator of the Anthracite Museum Complex; Gladi Kulp of the Alaska Historical Collections; Kathy A. Lafferty of the University of Kansas Libraries; Patricia Proscino Lusk of the Balch Institute for Ethnic Studies; Bernice M. Mooney, archivist of the Diocese of Salt Lake City; Barry Moreno and Jeffrey S. Dosik of the National Park Service; Steve Nielsen of the Minnesota Historical Society; John F. O'Brien of Cleveland's Irish Cultural Festival; Ríonach uí Ogáin of the Department of Irish Folklore, University College, Dublin; Richard Ogar of the Bancroft Library; Antonio Perez; Tony Pisani of the Museum of the City of New York; Anne Pollard, chair of the Trustees of the Pollard Library; John T. Ridge; Jim Rogers of the Irish-American Cultural Institute; John Sours, publisher of *Civil War* magazine; Eveard Stelfox, director of the Lane County Historical Museum; Kathey Swan and Eleanor M. Gehres of the Denver Public Library Western History Department; Janet Tearnen of the California Museum of Photography; Dan Warner of the Lawrence *Eagle Tribune;* Bonnie G. Wilson of the Minnesota Historical Society; David Wooters and Janice Madhu of the International Museum of Photography; Giles Wright of the New Jersey Historical Commission; Michael Wurtz and Michele Lacy of the Sharlot Hall Museum; and Diana Zimmerman of the Center for Migration Studies.

Finally, we owe more than we can express to Ita Donahue and her family. They graciously welcomed us to their home on what was probably the busiest weekend of the year—the celebration of St. Patrick's Day. We will never forget listening to Ita's account of her family history while we enjoyed an Irish tea. Guided by Ita, we toured Lawrence, Massachusetts, and saw the stark mills where her parents, and generations of Irish immigrants, worked. The high point of the day was a visit to the Ancient Order of Hibernians' Hall in Lawrence. To the sounds of bagpipers from Ireland practicing for the parade, history came alive as we examined the mementoes of Irish American history proudly displayed there. Slainte, Ita!

ABOUT THE AUTHORS

Dorothy and Thomas Hoobler have published more than 50 books for children and young adults, including *Margaret Mead: A Life in Science; Vietnam: Why We Fought; Showa: The Age of Hirohito;* and *Photographing History: The Career of Mathew Brady.* Their works have been honored by the Society for School Librarians International, the Library of Congress, the New York Public Library, the National Council for Social Studies, and *Best Books for Children,* among other organizations and publications. The Hooblers have also written several volumes of historical fiction for children, including *Frontier Diary, The Summer of Dreams,* and *Treasure in the Stream.* Dorothy Hoobler received her master's degree in American history from New York University and worked as a textbook editor before becoming a full-time freelance editor and writer. Thomas Hoobler received his master's degree in education from Xavier University and he previously worked as a teacher and textbook editor.